A GUIDE TO

SMALL BOAT
RADIO

By the same author

Astro Navigation by Pocket Computer
ISBN 0 229 11846 1

A GUIDE TO
SMALL BOAT RADIO

Mike Harris

ADLARD COLES NAUTICAL
London

Adlard Coles Nautical
an imprint of A & C Black (Publishers) Ltd
35 Bedford Row, London WC1R 4JH

First published in Great Britain by
Adlard Coles Nautical 1991

Copyright © Mike Harris

ISBN 0-7136-3436-7

Apart from any fair dealing for the purposes of research or private study, or criticism or review, as permitted under the Copyright, Designs and Patents Act, 1988, this publication may be reproduced, stored or transmitted, in any forms or by any means, only with the prior permission in writing of the publishers, or in the case of reprographic reproduction in accordance with the terms of licences issued by the Copyright Licensing Agency. Inquiries concerning reproduction outside those terms should be sent to the publishers at the address above.

A CIP catalogue record for this book is available from the British Library.

Typeset by Computape (Pickering) Ltd, N. Yorkshire, in Trump 10/12½

Printed and bound in Great Britain by
Hollen Street Press, Slough, Berks

Contents

Preface	vi
Acknowledgements	x
Terms and abbreviations	xi
1 Introduction	1
2 Communicating at sea – Licences and services	20
3 Amateur radio	32
4 Data modes	40
5 Radio equipment	63
6 Radio installation – Demands made on the boat	81
7 Antennas for use afloat	101

REFERENCE INFORMATION

I Codes and procedures	121
II Frequency lists	130
III Weather information	146
IV Technical data	157
Index	162

Preface

To be forewarned is to be forearmed.

Switch on a portable radio, tune in to some local station and, though it may just be pop music that you hear, day or night the broadcasts are always there. Switch off and the sounds stop, but the signals that brought them do not go away. Although we cannot hear or feel them, they form but a tiny fraction of the countless radio signals that surround every individual on the planet every moment of their lives. Of all the signals that might be heard some, like the pop music station, will be of local origin but others will be from more remote stations, as far away as the other side of the world. The range of information they contain is vast and rather like entering a library; there is something of value for everyone if only you know how to look.

Certainly we could be the best informed generation that ever lived. Just about any conceivable sort of information that we might require is there if we choose to look. However, take a trip in a small boat just a few miles off shore and this veneer of civilization quickly falls away and you may well become more isolated from humanity than an astronaut journeying to another planet. In spite of the increase in maritime traffic, there are great expanses of sea where it is still possible to spend months and see no other sign of human existence.

Of course, the desire to get away from it all could be a major reason for putting to sea in the first place and often I have heard people say that the last thing they want aboard their boat is a telephone. Joshua Slocum et al. did very well without it, so why is it necessary? If there is one lesson we can learn from the accounts of early pioneers it is that none of the mass of electrical gadgets fitted to so many of today's boats is essential. The sea still demands that sound sails, rigging, machinery, water tight integrity of the hull and skills of basic seamanship have first priority.

However, things are very definitely different today and it would be crass to ignore the wealth of information that is freely available by radio. To take advantage of a source of up-to-the-minute information on pilotage and weather information is simply good seamanship. To be able to find out about berthing charges, local customs, a changing political situation or of areas in which piracy is a problem, could save the budget, keep you out of prison or even avert something more serious. For those who fear that the presence of a radio aboard might bring with it an unwelcome invasion of privacy, then unlike the domestic telephone, there is always the off switch.

Marine VHF is a service already familiar to most small boat sailors and widely used as a source of navigational, safety and weather information, but its capabilities are strictly limited. Already in some areas it is over congested and has no place for routine chatter, but for transoceanic sailors its most serious limitation is its range. Essentially this is line of sight, or about 30 miles, and as you travel further from the coasts it becomes increasingly difficult to obtain even basic weather information, especially if your language abilities are limited.

The solution to these difficulties lies in radio equipment capable of operating on other frequencies, coupled with the knowledge of where and when to look for the information you need. These days such equipment need be neither large nor expensive and, if used thoughtfully, the power needed can easily be provided by a wind generator or solar panel. A high frequency transceiver the size of a shoe box can provide voice contacts over many thousands of miles, and it would seem that no boat is too small to have one aboard. A point well made by Tom Maclean, who installed one aboard his rowing boat and put it to good use whilst crossing the Atlantic in the summer of 1987.

ABOUT THIS BOOK

In writing this book I had two aims in mind. The first was to provide readers with an introduction to the variety of radio communications and information systems that can be achieved with a limited amount of equipment and within the confines of a small boat. This is the purpose of Chapters 1–7. I have mentioned a wide range of radio applications in terms that I hope readers with little technical background will find acceptable. Radio navigational aids are a topic that has not been included, the reason being that I felt that these form a large and separate subject which is at present undergoing many important but, as yet, uncertain changes. None the less the sections on propagation, interference and antennas are just as applicable to these radio systems as any others.

The second aim of the book is to provide directions towards sources of more detailed information and some basic reference material for those contemplating setting out on an extended voyage. This is the purpose of the Reference Information section of the book which, apart from including radio data and procedures, also includes frequency lists of stations likely to be of interest to mariners. In practice, most seasoned voyagers discover just a few stations that provide them with virtually all the information they need and these lists are compiled with this experience in mind. On the whole, these stations tend to be very well known and have often occupied the same spot on the frequency dial for years, but we are entering a time of change and, as a result of resolutions passed at the World Administrative Radio Conference in Geneva 1987, many changes will take place over the coming years. It seems likely that not all stations will switch immediately to their new allocations and even those that do may not give notice of their intentions. This makes the job of compiling long term frequency lists a rather uncertain task. Those that I have included are based upon the best available current information and it is inevitable that some changes will occur.

CAUTION

In all countries the use or even ownership of transmitting equipment is strictly controlled. In addition to issuing operator authorizations, national or local legislation may regulate the type of equipment that can be used, its maximum power, the mode of transmission and the type of messages that can be passed. Unauthorized transmission or even unintended interference is likely to bring official action. Penalties can be severe and may include confiscation of the equipment, heavy fines, the impounding of your vessel or imprisonment.

Regulations controlling the use of radio equipment vary considerably between countries. Officials usually ignore equipment installed aboard visiting foreign boats, but the same attitude may not be taken with equipment that is brought ashore. Listening in on certain locally sensitive frequencies or simply having the equipment that would enable one to do so may be forbidden.

Ignorance of the law is no protection. So when visiting foreign countries do take the trouble to obtain all authorizations and make yourself aware of legislation relating to radio use. An advance enquiry to the national radio regulatory organization, radio society, or in the last resort local port officials, should bring details of what is required.

Mike Harris
Gibraltar, 1991

Acknowledgements

In writing this book, many people have provided me with much help and encouragement. First, thanks go to my wife Di, for preparing the artwork, checking the text and for her continued support while I was writing it. My thanks also go to Bob Laws, George Wheatley, David Jolly and Jimmy Griffin for their helpful and constructive criticisms on the text. Also to:

British Telecom International
Portishead Radio
GIBTEL, Gibraltar
Interactive Systems, Gibraltar
South Midlands Communications
ICS Electronics Ltd
ICOM (UK) Ltd
The Radio Amateur Licensing Unit

Communicating is what radio is all about: last, but by no means least, I am grateful for the help I have received from radio amateurs, in particular those stalwart controllers of the UK maritime mobile net. Others are unfortunately too numerous to mention. They come from many different countries, some ashore, some afloat, some I may never meet, but without them I would never have got started.

Terms and abbreviations

Abbreviation	Meaning
AMTOR/SITOR	Amateur and commercial implementations of telex on radio (TOR)
AC	Alternating current
AF	Audio frequency
AM	Amplitude modulation
AFSK	Audio frequency shift keying
ARQ	Automated repeat request
ITA	International telegraph alphabet
BIT	A single unit of binary data
BYTE	The collection of bits that make up a binary word
BAUD	A measure of the rate of transfer of binary messages 1 bit/second = 1 baud for most purposes
CB	Citizen band
CEPT	European Conference of Postal Telecommunication Administrations
CES	Coast earth station
CW	Carrier wave
DC	Direct current
DSC	Digital selective calling
DTI	Department of Trade and Industry
EEC	European Economic Community
EMC	Electromagnetic compatibility
EPIRB	Emergency position indicating radio beacon
FAX	Facsimile
FEC	Forward error correction
FM	Frequency modulation

FSK	Frequency shift keying
GMT	Greenwich Mean Time (see Z below)
HF	High frequency (3 MHz to 30 MHz)
INMARSAT	International Maritime Satellite Organization
LAMTOR	AMTOR in the listening mode (see AMTOR)
LSB	Lower side band
LF	Low frequency (30 kHz to 300 kHz)
IARU	International Amateur Radio Union
ITU	International Telecommunications Union
LW	Long wave (i.e. Low frequency see LF)
MF	Medium frequency (300 kHz to 3 MHz)
MW	Medium wave (i.e. Medium frequency see MF)
NAVTEX	Navigational Information transmitted as FEC, SITOR
PEP	Peak envelope power – a measure of SSB transmitter output power
RF	Radio frequency
RALU	Radio Amateur Licensing Unit
RDF	Radio direction finding
RST	Readability, strength and tone (code)
RT	Radio telephone
RTTY	Radio teletype
RYA	Royal Yachting Association
SES	Ship earth station
SID	Sudden ionospheric disturbance
SSB	Single side band
SWR	Standing wave ratio
SW	Short wave (i.e. High frequency see HF)
TNC	Terminal node controller
TOR	Telex on radio
TVI	Television interference
USB	Upper side band
UTC	Universal coordinated time (see Z below)
VHF	Very high frequency (30 MHz to 300 MHz)
UHF	Ultra high frequency (300 MHz to 3000 MHz)
VLF	Very low frequency (3 kHz to 30 kHz)
WT	Wireless telegraphy
WARC	World Administrative Radio Conference
WEFAX	Weather fax
YTD	Yacht telephone debit
Z	Zulu time, which for the purpose of this book is synonymous with GMT or UTC

1. Introduction

Over the last few decades, advances in communications technology have come thick and fast. Semiconductor technology has largely replaced valves, making equipment smaller and less hungry for power. Integrated circuits and microprocessor technology have brought further savings in space and made equipment more friendly to its users. As the technology has matured, so reliability has increased and the current generation of marine radio equipment would hardly be recognizable as performing the same function as that installed aboard vessels of 25 years ago. In spite of this continuous technical innovation and improvements in hardware, some less obvious aspects of radio use have hardly changed at all. The physical principles which make radio transmissions possible are, of course, unchanging but perhaps more surprising is that some radio techniques that are taken for granted today were first developed years before radio was discovered.

We begin this chapter with a look at telegraph systems, and the kind of codes that were used for long-distance communication some 100 years before radio appeared, but which remain in use today and are expected to remain in use into the next century. Next, we look at some early wireless experiments and some characteristics of radio signals. Finally, the chapter ends with a description of different types of radio emissions used in marine communications today.

EARLY TELEGRAPHY

The idea of using electric signals to communicate over long distances was one of the earliest applications of electricity. Like many great inventions the telegraph was developed by two groups of people working independently of each other, but it was the

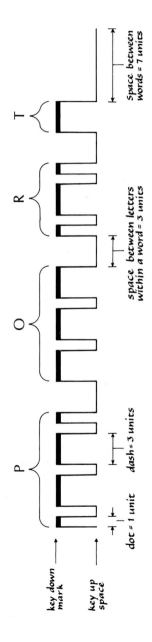

Fig. 1.1 Time relationships in Morse code. The code may be sent faster or slower but, to remain readable, the time relationships must not be altered.

American Samuel Morse who, in the 1830s, produced the system that was to stand the test of time. Unlike the British idea, his system used a single pair of wires to connect sending stations to receivers, and a current sent through this circuit could be detected at the receiving station, though it would be much reduced if the distance was too great. With the equipment of the time it was only practically possible to say whether current in the circuit was turned on or off. In using this for a communications system, there was the problem of encoding numbers and alphabetic characters as a sequence of *on/off* (later known as mark/space).

The Morse code

The most well-known feature of Morse's code were the two characters that he used, i.e. the dot and dash. These corresponded to a short or a long period of current flow and later, when the code was used over radio, they became referred to as *dit* and *dah* because of the kind of sound that was made. By combining these characters with different space periods he was able to encode the whole alphabet and numbers 0 to 9. Over the years there have been slight changes to the code but the form in which it is used today remains essentially the same (see p. 12).

For Morse to be received correctly, accurate timing is vital and though the exact duration of dot is not so important, its relationship to the period of the dash is. Also important is the delay between dot/dash characters and the delay between letters and between words. Fig. 1.1 shows the ideal relationship.

On telegraph systems Morse code remained in use for over 100 years and during this time demands for the service increased substantially. More cables were erected and networks emerged. Intercontinental cables were laid and the need for a more efficient means of sending messages soon became apparent.

The later part of the 19th century was a great time of mechanical ingenuity and there seemed almost no limit to the kind of work that inventors would build machines to carry out. Machines were already sending and receiving Morse from punched paper tapes but the basic nature of the code restricted further progress in this direction. The main difficulty was that different letters contain different numbers of dots and dashes, so making some quicker to send than others. This created problems with receiving machines and very precise timing was needed to know exactly when one letter had finished and another was about to begin.

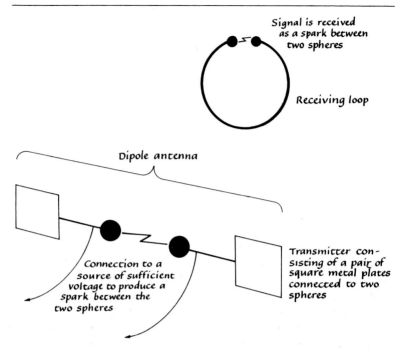

Fig. 1.2 Hertz's transmitter and receiver.

Teleprinter codes

The machine-readable telegraph code that eventually replaced Morse was quite different. It had no separate dots or dashes and each mark/space unit was of equal length. Individual letters were encoded as combinations of five such units, or *bits* as they are known today. With such a code, 32 different 5-bit combinations are possible, but in current versions two of these are used to indicate letter or figure shifts. This makes it possible to encode the whole alphabet, along with numbers 0 to 9, and a few punctuation and control codes. (See ITA No. 2 code p. 122.)

By 1920 most wire telegraph systems had been converted to use the new codes. Using a technology borrowed from the printing industry, typewriter-style machines were used to transpose messages into five-unit hole patterns in a paper tape and a further machine then converted these into the electrical code. At the receiving station the whole process was reversed but was again fully automated.

All this new development did not mean that the use of Morse code was about to die out. Far from it, for with the advent of radio a whole new future lay in store. Only the simplest of radio equipment is needed to send Morse messages. Even today it is effective over long distances, particularly when little power is available and radio conditions are poor.

EARLY WIRELESS

To describe a radio as a wireless is something of a misnomer. Take a look inside any set and you are likely to see plenty of wires but at the end of the 19th century, when all long-distance telegraphic messages were sent along wires, the term had more meaning. The Italian Guglielmo Marconi is the man usually credited with its invention, but his genius lay in his abilities as a far-sighted practical innovator rather than as a theoretician. The foundations for his achievements were laid in the previous century, when James Clark Maxwell published his theoretical description of electromagnetic radiation. This was a mathematical theory and proposed the existence of a range of different types of electromagnetic radiation of which visible light formed a small part.

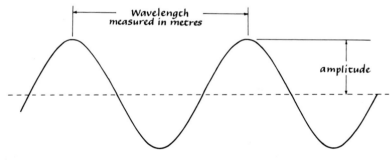

Frequency = No. of wavelengths that pass per second = $\dfrac{c}{\text{wavelength}}$

Where c = Speed of light in free space
= 300×10^6 m per sec

Radio waves can be described in terms of wavelength or frequency

Units:
1 Hz = 1 hertz = 1 wavelength per second
1 kHz = 1,000 Hz = 1 kilohertz
1 MHz = 1,000,000 Hz = 1 megahertz

Frequency (in MHz) = $\dfrac{300}{\text{wavelength (in metres)}}$

Wavelength (in metres) = $\dfrac{300}{\text{frequency (in MHz)}}$

Fig. 1.3 Relationships between frequency and wavelength in radio waves.

Maxwell also predicted that such waves filled the universe and could travel great distances. Unfortunately, these ideas were only believed by a handful of his contemporaries. Today these theories are well established and accepted by physicists almost without question.

Sadly, Maxwell did not live to see the experimental proof of his work and died ten years before Heinreich Hertz staged the first demonstration of wireless transmission. His apparatus was simple (see Fig. 1.2) and the ranges he achieved were only a few metres but they were enough for the far-sighted Marconi to realize the possibilities for a communication system. By switching the spark generator on and off (i.e. keying it) with Morse, he was able to send messages. Furthermore, he experimented extensively with antennas, reflectors and improved signal detectors and managed to increase the range from a few metres to several miles.

Surprising though it may seem today, when Marconi came to England in 1896 he found it very hard to convince the public that there could be any practical value in wireless communications. None the less he continued giving lectures and demonstrations, many of which were carried out at sea where there could be no possibility for any other method of communication. Then, in 1899, he successfully equipped a pair of US ships with radio equipment to report on the progress of the America's Cup and the achievement aroused interest and excitement world-wide.

The limitations of spark transmitters

A characteristic of electric sparks is that they produce radio waves across a broad spectrum of frequencies. As an example, the unsuppressed spark ignition systems on outboard motor or car engines are not designed as transmitters, yet the radio interference they produce can blot out reception of both television and radio stations over a wide area. For this reason, the use of spark transmitters is now strictly forbidden, and even for Marconi, these broad-spectrum transmissions meant that different stations were unable to operate simultaneously without causing interference to each other. As usual, the problem did not defeat him and, in 1900, he filed a patent for improvements for tuning transmitters to different frequencies so that they could work independently.

In 1901 some of the best-informed mathematicians were of the opinion that radio contacts were only possible over distances of 100 to 200 miles, and at greater distances the passage of radio

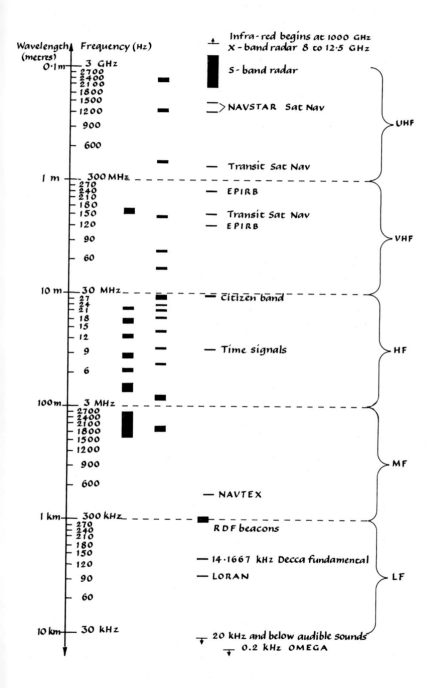

Fig. 1.4 The radio spectrum.

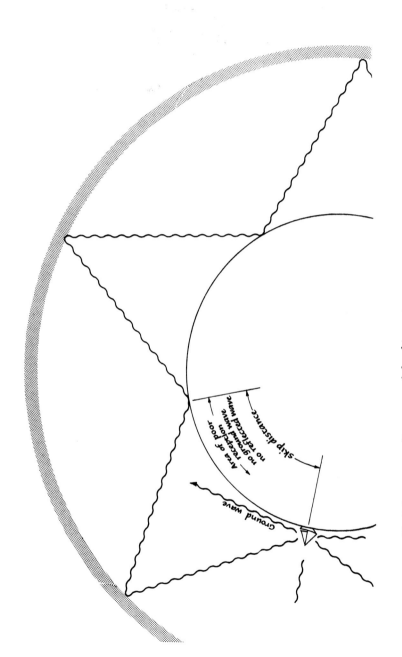

Fig. 1.5 Refraction of high frequency radio waves and the skip zone.

waves would be blocked by the curvature of the earth. In spite of this advice Marconi was undiscouraged and continued to push for ever greater distances. Towards the end of the year he was even foolhardy enough to make an attempt at the first transatlantic crossing by radio – it was an unreserved success.

THE RADIO SPECTRUM FOR SMALL BOATS

Whereas the visible light spectrum covers the familiar colours, red to violet, the radio spectrum covers an even greater range of longer wavelengths that we are unable to see or feel. At one end there are very low frequencies (VLFs) which, if they were sound waves, would be within the range of normal human hearing, and at the other end there are extremely high frequencies approaching that of infra-red light. Though this is only a part of the whole electromagnetic spectrum, it is a truly vast range of radiations and, maybe because we have no physiological means of detecting them, it can be difficult to appreciate the scale of things. Those frequencies at the top end of the range are around ten thousand million times greater than those at the bottom. Or put another way, it spans a range of wavelengths from a few centimetres to some 15 nautical miles. As far as marine users are concerned, there are important uses occurring right across this scale, from low frequency position fixing systems at one end to radar at the other. Fig. 1.4 shows the radio spectrum with frequencies of special interest to mariners.

Propagation of different frequencies

Radio frequencies of VHF and above tend, rather like light, to travel in straight lines. They can travel great distances into space but their terrestrial range is restricted by the curvature of the earth. On the other hand, waves of medium and lower frequencies travel large distances, close to the surface of the earth, but the atmosphere tends to prevent them from leaving. Between these extremes lies the HF section of the spectrum and here propagation characteristics bear some similarities to those of VHF and MF. However, with HF propagation the earth's ionosphere has the most profound influence and can be responsible for signals being heard with amazing clarity 5000 miles away or being totally blocked out after 50.

IONOSPHERIC REFRACTION

The ionosphere consists of a series of layers of ionized gases in the upper atmosphere. These surround the earth and have the property of being able to bend certain radio frequencies, causing them to be returned to earth. In this way they can be received at very great distances from the transmitter, well beyond the optical line of sight. By repeated reflection between earth and ionosphere such signals may even be received on opposite parts of the globe (see Fig. 1.5).

Within the upper atmosphere ions are formed when gas molecules lose electrons. The energy for the process comes from collisions with X, ultraviolet or other sun radiations, but the transition is reversible and the freed electrons are able to combine with ions to reform gas molecules. At any particular time the proportions of molecules in an ionized state (the degree of ionization) will depend upon the prevailing level of radiation and, as a result, more ionization will occur over those parts of the earth that are in daylight. In this way, radio propagation tends to follow a daily pattern so that you may notice better radio reception from stations that lie towards the direction of the sun, rather than from those in darkness.

In addition to these diurnal patterns there are seasonal changes, as well as an important 11-year cycle following the sun-spot activity. Sun spots usually accompany high levels of solar radiation which, in turn, lead to increased ion production. Also, because higher levels of ionization are required to refract the higher frequencies, these are times when frequencies of 20 MHz to 30 MHz can be used to cover large distances – a point that does not go unnoticed amongst Citizen Band radio enthusiasts.

LAYERS WITHIN THE IONOSPHERE

The layering shown in Fig. 1.6 is typical of that found during the daytime. During the night, when the sun's radiations are absent, the D and E layers tend to disappear whereas the F1 and F2 layers merge and form a single F layer, which is mainly responsible for long-distance night propagation. Ion densities and heights of all layers can change, depending upon solar conditions, but each has its own characteristic effect upon radio propagation.

F, F1 and F2 layers
During the day the highest layer is the F2, which also has the

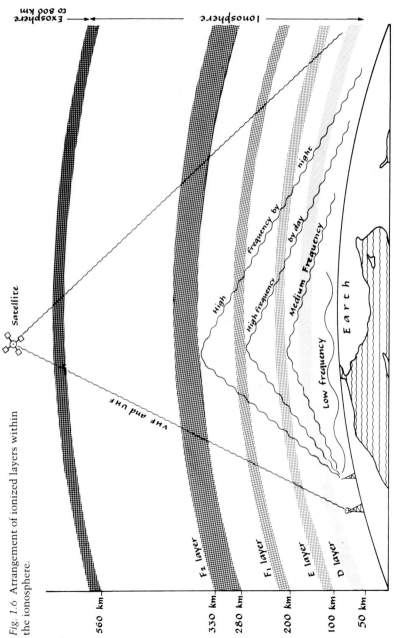

Fig. 1.6 Arrangement of ionized layers within the ionosphere.

highest ion density and so is capable of returning frequencies around 22 MHz, near to the top end of the HF range. This is useful in covering distances of 2500 miles or more.

F1 layer
When the F1 is not combined with the F2, it is less highly ionized and is capable of returning frequencies of the mid-HF and is used for distances of around 2000 miles.

E layer
This layer has an even lower ion density but is effective in returning lower frequency signals. These are useful in covering distances of a few hundred to 1500 miles.

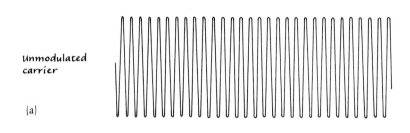

Fig. 1.7 Keying a carrier with Morse.

D layer

The D layer has the lowest ion density and tends to absorb rather than refract radio signals. Higher ion densities in this layer tend to produce fading on HF bands.

DEEP FADING

Sometimes, following solar flares, the sun produces abnormal amounts of UV and X radiation which lead to greatly increased levels of ionization in the lower D layers. This has the effect of absorbing rather than reflecting radio waves and during these times HF radio communications may be totally lost. For a few minutes, or maybe hours, it is possible to tune across the bands and hear nothing other than sounds like distant bacon frying. The effect is known as a 'Sudden ionospheric disturbance' (SID) or 'Dellinger' fade out. A few days after the SID another kind of intense fading, the 'ionospheric storm' may follow. This is thought to be caused by the arrival of slower moving particles from the same flare.

FINDING OUT WHICH FREQUENCIES ARE AVAILABLE

In addition to diurnal ionospheric variations, there are seasonal changes and other effects which make it difficult to predict the exact areas and distances over which particular frequencies will be effective. The Radio Society of Great Britain publishes monthly predictions, and there are computer programs that can help with the task, but it is far from an exact science. For individuals without these facilities there are few hard and fast rules, but as a guide:

1. Listen to your radio.
   Listening in to beacons and traffic lists will tell you which bands are open and, by listening regularly, you will be able to build up your own picture of propagation changes.
2. If you can hear them, they will hear you.
   Of course there are many reasons why this rule should not work but it can be a guide in choosing a frequency on which to make a contact.
3. The table overleaf provides a rough guide to the ranges achievable with various frequencies (actual ranges will vary with conditions):

Frequency band (MHz)	Daytime range	Night-time range
1.5 – 3	Ground wave only	1000 miles
3 – 6	Ground wave only	1500 miles
6 – 10	600 miles	2000 miles
10 – 16	1800 miles	World-wide in the direction of the sun.
16 – 23	3000 miles	World-wide in the direction of the sun until 20.00 local time.
22 – 30	May be world-wide but depends upon ionization.	Little sky wave reflection after sunset.

In general, low frequencies are better for longer distance contacts at night.

4. The BBC Publication *London Calling* is primarily intended to give details of forthcoming programs. It also includes a table showing which frequencies can be expected to give best reception at certain times in particular parts of the world. This, coupled with a knowledge of the location of transmitters (also available from the BBC), can provide a useful source of HF prediction.

TYPES OF MODULATION (MODES)

When a radio receiver is tuned to a frequency close to that of a radio frequency carrier wave (of the type shown in Fig. 1.3 and 1.7(a)) the signal may be heard as an audible tone with a pitch equal to the difference between the frequency of the carrier and that to which the radio is tuned. This is fine as far as it goes, but as the signal is nothing more than an audio tone it is not able to carry any useful information. To make it do this involves changing its shape, i.e. modulating it in some way, and there are a variety of ways in which this may be carried out. Some of the more widely used methods are listed in Reference Information, section IV, on the designation of radio emissions. Of these methods, by far the simplest is to key the carrier with Morse, as in Fig. 1.7(b) (i.e. A1A). Morse transmissions are sometimes referred to as WT (wireless telegraphy) or CW (carrier wave).

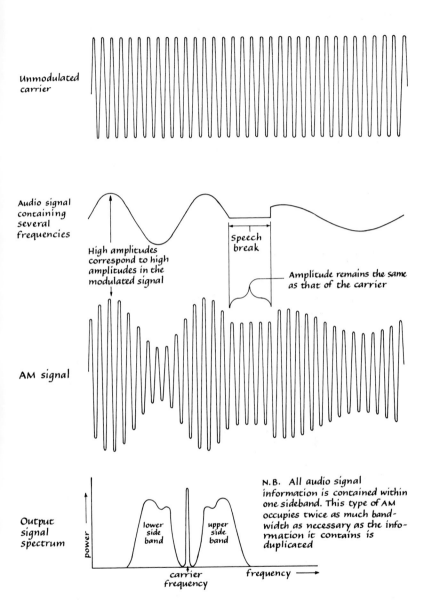

Fig. 1.8 Amplitude modulation.

Telephony

Speech sounds can be converted into electrical audio signals with a microphone but, by and large, their frequencies are very much less than those of the radio frequencies used for communications. Within a radio transmitter the purpose of the modulator section is to combine these audio signals with the radio frequency carrier, and there are several different ways in which this can be carried out.

AMPLITUDE MODULATION – AM (A1A)

This is the oldest method of speech modulation, shown diagrammatically in Fig. 1.8.

These days, quite apart from maritime applications, there are broadcast stations, aeronautical mobiles, news services and an enormous number of other users all competing for sections of the HF bands. Whilst a sharply tuned carrier wave occupies only a single spot frequency on the bands, once it is modulated in some way it inevitably spreads to adjacent frequencies. Many HF bands are already over used and any measures that can be taken to limit this spread (or bandwidth) will help ease the congestion. The following table gives a list of the bandwidths associated with different types of transmissions:

Transmission type	Bandwidth
Morse telegraphy	>100 Hz
Radio telex NAVTEX	100 Hz
WEFAX	2 kHz
SSB communications	3 kHz
Music	15 kHz
High-speed computer data	<1 MHz
Colour television (625 line)	5.5 MHz

As far as speech is concerned, the normal audio range is between 20 Hz and 16 kHz, but perfectly good conversations can be held by only accepting those that fall within the range 2.4 kHz to 3 kHz. This does not give what might be called hi-fi reproduction and would be unsuitable for music, but by using this restricted range the bandwidth of the modulated signal is similarly reduced.

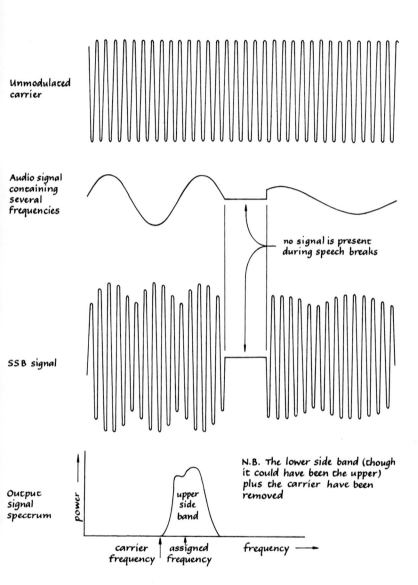

Fig. 1.9 Single side band modulation.

SINGLE SIDE BAND – SSB (J3E)

Today, amplitude modulation is almost never used for marine communications and has been replaced by single side band (Fig. 1.9). This type of modulation has the advantage of a narrower bandwidth and also makes more effective use of transmitter power. The disadvantage is that this makes SSB transmitters rather more complex than those for AM.

In essence, most SSB transmitters work by producing a conventional AM signal, then filter out part or all of the carrier wave along with all frequencies to one side of it. With upper side band SSB only frequencies above the carrier are transmitted; with lower side band SSB only the lower frequencies are transmitted.

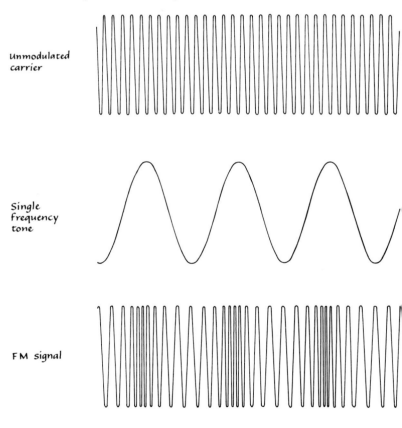

Fig. 1.10 Frequency modulation.

FREQUENCY MODULATION – FM (F3E)

With frequency modulation the amplitude of the transmitted signal does not change but its frequency is varied by the audio modulating signal. Fig. 1.10 shows how.

When compared with AM or SSB, an advantage of FM is that signals are usually received with lower background noise levels. Set against this is the disadvantage that FM requires a wider bandwidth. For this reason it is generally only used on VHF and higher frequencies where more space is available.

2. Communicating at sea – Licences and services

At home ashore we can pick up the telephone and speak to anyone anywhere in the world, just so long as they also have a phone, but to what extent is it practically or financially feasible to use this kind of communication in a small boat at sea? The sections of the MF, HF and VHF bands that are allocated for marine use, are intended to provide all vessels, large and small, with the means of communicating with each other and with telephone subscribers anywhere in the world. In this chapter we look at the kinds of radio services that can be found on these bands and begin by reviewing the types of licences and qualifications that we need to make use of them and services to be found on them. Though the licensing requirements described here are those in force within the UK, many other countries adopt similar conditions.

LICENSING REQUIREMENTS

The use of radio equipment aboard boats is covered by several different types of licence, but they broadly fall into three groups, i.e. operator qualification/authorization, licences for the vessel or particular items of its equipment and equipment type approvals.

Operator qualifications and authorization

The four types of radio operator qualifications/licences most likely to interest small boat users are:

1. Restricted (VHF only) RT Certificate.
2. Restricted RT Certificate.

3. Citizen Band Radio Licence.
4. Amateur Radio Operator Licence.

Of these, amateur radio is covered in Chapter 3, but the first two are professional qualifications covering the use of radio telephones at sea. To obtain an RT certificate involves passing an examination, the reason being that the marine bands are used for essential services, including the control of search and rescue operations. When these are taking place it is vital that all users understand what is expected of them and, above all, do not cause interference. In the normal course of events there is also the need for users of the system to adopt efficient procedures and not to block communication channels for an unreasonable length of time or cause offence to other users.

EXAMINATIONS

The examinations are by no means onerous and many users will find the slightly simpler VHF only certificate sufficient for their needs. For those expecting to venture off shore and needing longer range communications, the standard Restricted RT certificate is more appropriate. In both cases candidates are expected to show:

(a) A practical knowledge of radio-telephone operation and procedure.
(b) An ability to send correctly and to receive correctly by radio telephone.
(c) General knowledge of the regulations applying to radio telephone communications and specifically to that part of the regulations relating to the safety of life.

Details of examination centres can be obtained from the Marine Examinations Group (in the case of the Restricted RT Certificate) or the Royal Yachting Association (for the Restricted [VHF only] Certificate). See contact address list.

Candidates are required to sign a declaration saying that they will preserve the secrecy of any radio correspondence that they may hear. On passing the exam the Department of Trade and Industry (DTI) usually issue a Certificate of Competence and also a Certificate of Authority to Operate. Both of these normally remain valid for life, though, if there are sufficient grounds, the Authority to Operate may be suspended.

VESSEL/EQUIPMENT LICENCES

These licences are renewable and cover both the vessel and its radio equipment. There are two types:

1. *Ship Radio (VHF) Licence*
   This entitles the licensee to install or establish a VHF maritime radio for use on the international VHF maritime frequencies (see Reference Information, section II). The licence also covers all radio receivers (including Decca, NAVTEX, Loran, Satnav, RDF, etc.), radar and EPIRBs.
2. *The Full Ship Radio Licence*
   This entitles the installation of all maritime radio equipment, including satellite communications equipment (e.g. ship earth stations, see p. 30) on international maritime frequencies. This licence also covers low-power on-board communications equipment.

Both licences impose a number of conditions such as:

(a) The frequencies to be used.
(b) The maximum permitted output power.
(c) The type of modulation.
(d) Only permits the station to be operated under the control of a holder of a Certificate of Competance and Authority to Operate.

This last condition does not preclude non-certificate holders from using radio telephones, but allows them to do so only if they are under the control of a qualified operator.

These licences may also authorize the use of private channels such as channel M (i.e. 157.850 MHz) which, within the UK, is used for communications with marinas or yacht clubs. (In different countries this frequency is used for other purposes and may be paired with another frequency to form half of a duplex channel.)

CALL SIGNS

Call signs are issued along with the first application for a ship licence and these remain with the vessel indefinitely even after a change of ownership. Their purpose is to identify stations so that sources of interference may be located and vessels correctly billed for coastal radio station services. For this purpose, and certain transmissions eg. Telex, the call sign may also be combined with other details on the ship licence, such as a SELCAL number.

THE SHIP RADIO (TRANSPORTABLE) LICENCE

This is a VHF licence intended to cover hand-held transceivers and unlike the standard ship radio (VHF) licence, the transportable licence covers the radio equipment and its owner, rather than the vessel and its owner. The advantage of the arrangement is that it allows a hand-held transceiver to be used on different vessels under one licence. A call sign is not issued with the licence but holders have access to all marine VHF channels and are able to make telephone link calls. Billing can be arranged through the Yacht Telephone Debit system (YTD), or by quoting the licence reference number, though these facilities are not available outside the UK.

CITIZEN BAND (CB) RADIO/LICENCE

Unlike marine or amateur radio, no examination passes are needed to use CB frequencies. A licence is still required but formalities are minimal and for millions of people it has been an easy and attractive way of getting on the air. CB radio operates at frequencies around 27 MHz, and on the odd occasions when conditions are favourable quite large distances can be covered, though for most purposes ranges are unlikely to extend much beyond line of sight. In the United States and many other countries where CB is permitted it is transmitted as an AM signal, but in the UK and some western European countries FM is the standard, and the use of imported AM equipment is forbidden.

There are many groups of people for whom it fulfils a useful role, the best known being truck drivers, but in many towns and cities it is unfortunate that its popularity has led to overcrowding of the bands. Far more serious is the emergence of a small minority of mindless, moronic radio users, who devote their time to causing annoyance and disrupting the system. In areas in which they operate, these people effectively limit the usefulness of CB radio.

At sea, users are in a more advantageous position. Once away from population centres, activities on the bands decrease and, as a means of chatting to friends, CB has a useful part to play. There are far fewer restrictions attached to the use of CB than with the marine band frequencies. You can talk about almost anything (no soliciting goods or services, transmission of offensive language or music), and there are no restrictions on talking to shore-based stations. In some parts of the Caribbean, CB is used extensively for inter-island communication.

Caution
Whilst CB can be useful at sea, it would be wrong to expect it to act as a substitute for marine band VHF. It is not monitored by ships or coastguards and they may not be equipped to receive it.

TELEVISION BROADCAST RECEIVING LICENCE

Licensing requirements for television sets aboard yachts are similar to those applied to homes ashore.

EQUIPMENT-TYPE APPROVALS

Any manufacturer of marine band transceivers must ensure that they conform to set standards. These define the frequency ranges that the equipment must cover, its stability, the method of changing frequency and its power output. Only when a particular set meets these standards is it given official 'type approval'. The Department of Trade will only issue a ship licence for type-approved equipment.

THE MARINE BANDS

Marine VHF

A characteristic of marine VHF is that it usually provides reliable communications with good quality speech and little background noise. This is because VHF is independent of the vagaries of ionospheric propagations and uses frequency modulation. These transceivers first became popular on small boats some 20 years ago when lower-priced sets first began to appear on the market. Since then marine VHF has rapidly increased in popularity and now plays an essential part in most search and rescue operations.

CHANNELS

The part of the radio spectrum used for marine VHF occupies the section between 156 MHz and 174 MHz, but to make operating procedures easier this has been subdivided into numbered channels. The most important is channel 16 (on 156.8 MHz), which has been designed as an international distress, safety and calling frequency. For safety reasons all stations at sea are encouraged to monitor this channel but, when employed for calling, users are required to continue their conversation on another channel once they have established contact. Other channels are allocated for specific purposes, e.g. intership communications, port operations, public correspondence and (in the UK) yacht safety. These are listed in section II of Reference Information.

On looking down the list you will notice that some channels, e.g. 16, are transmitted and received on the same frequency. These are known as *simplex* channels and in practice this means that a pair of people in conversation must take turns in speaking and use something like the word *over* to indicate when it is the other person's turn. Other channels, such as those used for public correspondence, are *duplex* channels and are transmitted on one frequency and received on another. In principle, this makes it possible to hold a normal telephone-like conversation where both people may speak at the same time. In practice, many sets are not able to transmit and receive at the same time and so are incapable of full duplex working. To overcome this problem most smaller sets are pre-programmed to switch between the receive and transmit frequency as the microphone push-to-talk switch is pressed. This is described as *semi-duplex* operation.

When listening into another ship station talking to a coastal station on a duplex channel, your set will be tuned to receive on a frequency which is different from the one on which the other ship station is transmitting. As a result you will not hear the ship, though you may hear the coast station. This characteristic makes intership working on duplex channels impossible.

VHF RANGE

As was mentioned in the last chapter, VHF range is essentially line of sight and though with directional antennas and under certain conditions this may be considerably extended, a restricted range can be a distinct advantage. These days, when there are so many users making demands on the VHF system, it does allow more stations to make use of the same frequencies without interfering with each other; more people can use the system than if the same services were carried out on, say, HF with world coverage.

However, responsible use of VHF still requires consideration for the needs of others. Once you have established contact with another station, it is good idea to reduce transmitter power to the minimum required to maintain the conversation. There could be others using the same frequency and though you may not hear them, this procedure will help reduce interference and makes a contribution to saving battery power.

THE 'CAPTURE' EFFECT

It is worth bearing in mind that, on VHF, signals received from a strong station have the effect of blanking out those from other weaker stations. So if your signals are strong, and you are in contact with another station with a similar signal strength, it is likely that neither of you will be able to hear a third, weaker station that may be trying to break in on your conversation.

LIMITATIONS OF THE VHF SYSTEM

Within the UK at least, the coastguard policy of phasing out many small, manned observation posts in favour of centralized radio stations has meant that in an emergency the importance of VHF now rivals that of traditional visual signals such as flares. However, in some areas the effectiveness of VHF communications has been seriously threatened by its own success. When large numbers of operators are demanding use of the system, there is an increased possibility of urgent or distress traffic going unheard, but several steps have been taken to preserve the integrity of the system under pressure. More channels have been added to the existing band allocation and calling for coastal radio stations is now carried out on their working channels instead of channel 16. However, in spite of this, the problem is still serious and any steps taken to alleviate congestion will also help reduce the possibility of an urgent or distress call being ignored.

One of the most widespread abuses of marine VHF is its use for general chat and, as important as it may be, there are no channels where you can talk about the price of fish or discuss local scandal. Elimination of this would undoubtedly help free more channels and could be achieved by the more widespread use of Citizen Band radio for these conversations.

Marine MF and HF

Marine VHF is fine when you are within 30 or 40 miles of a coastal radio station, but at greater distances high and medium frequencies provide the simplest means of keeping in touch. Like marine VHF, these bands are also subdivided into channels (see Reference Information, section II) and are similarly allocated to specific purposes, such as distress, safety and calling, intership use, telephone connections, telex and WT.

The essential difference between MF/HF and VHF is that results on the former are very much affected by ionospheric conditions and on any particular occasion some frequency bands will be more effective than others. In fact, there will be times when some bands are totally unusable. This means that, to get the best out of the

Photo 2.1 A WT control desk at Portishead Radio. Though Morse is slowly being phased out, many ships are still equipped for transmitting and receiving it and so it is likely to remain in use well into the next century.

Portishead radio is one of the largest commercial radio stations in the Western world and handles vast amounts of long-distance HF and MF traffic. Apart from WT, the station also provides an extensive range of RT and telex services and, though much of the routine work is handled by computer, radio officers are able to provide assistance when necessary. (*Courtesy of British Telecom plc*) system, users need to know something of ionospheric propagation in order to make a judgement on which frequencies to use.

Coastal radio stations

Just about every country that maintains a maritime interest has established its own system of commercial radio stations, the function of which is to handle communication traffic between the country's telephone/telex network and ships at sea. The range of services on offer from these stations varies considerably. Some may only work on VHF for limited hours whereas others may provide full-time services on MF and HF, with facilities for telex, WT and automatic handling of calls. Their services may also include transmission of weather information, navigational warnings and a responsibility for coordinating search and rescue work.

Comprehensive details of marine radio station schedules, frequencies and services can be found in the *Admiralty Lists of Radio Signals* and other publications. (See the book list, p. 159.)

Satellite communications

MF and HF have provided the mainstay of long distance marine communications for many years, but there can be long periods when electromagnetic disturbances within the ionosphere make this type of radio contact impossible. Another difficulty is that these services have become so heavily used that people who want to make telephone calls are often required to wait their turn. Even when they get through, atmospheric conditions sometimes make the contact difficult. Satellite systems have largely overcome these problems and interference-free calls can now be made both to and from vessels at sea with the same privacy and convenience that you would expect on land.

At present the only non-military satellite communication system for mobile users is provided by the International Maritime Satellite Organization (INMARSAT) – see Fig. 2.1. This consists of representatives from some 48 prominent maritime countries and provides telephone, telex, data exchange, facsimile and distress and safety communications services for offshore communities and vessels at sea. The three key elements in satellite communication systems are:

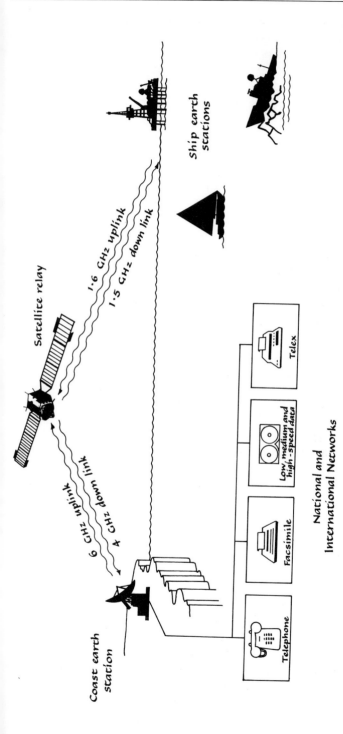

Fig. 2.1 INMARSAT satellite and coastal stations provide the interface between mobile stations at sea and international telephone and telex networks.

1. *The ship earth station (SES)*
   Usually, this refers to a mobile station, such as a vessel at sea, but could be an oil platform or other offshore community. These transmit signals to satellites on a frequency of 1.6 GHz and receive on 1.5 GHz.
2. *The space segment (i.e. satellite relay).*
   Over the equator a group of geostationary satellites receive messages from SESs on 1.6 GHz and retransmit them to coastal stations on 4 GHz. They also receive signals from coastal stations in 6 GHz and retransmit them to ship stations on 1.5 GHz. Satellites have been deployed to give coverage to most parts of the globe but the service at present does not extend to extreme polar regions.
3. *The coast earth station (CES)*
   These are land-based stations and provide the connections between satellites and international networks for telephone, telex, fax, etc. They should not be confused with coastal radio stations, which handle MF, HF and VHF traffic.

HOW CAN SATELLITE SYSTEMS BE OF USE ABOARD SMALL BOATS?

Satellite systems now have a well-established role to play in small boat navigation and EPIRB-initiated search and rescue operations. No doubt the future will bring a continued expansion of these services but for the present, as far as non emergency traffic is concerned, INMARSAT provides mariners with two separate communications systems.

INMARSAT Standard A

This is a comprehensive satellite service providing phone, telex, fax and data communications. The size of the radome used to operate the system restricts its application to large vessels only, e.g. tankers, container vessels, passenger ships and larger pleasure yachts.

INMARSAT Standard C

Equipment required to operate this system includes a small antenna, personal computer and electronics package. At the time of writing it is claimed to be the world's smallest, most portable commercial satellite communication system – small enough to fit into a shopping bag.

Standard C is a low-cost, text-only service for sending and receiving telex, electronic mail, computer-type graphics and data. In addition to these two-way communications there is also a receive-only service that can be used to circulate information to large numbers of vessels in a given area. An example of this is 'safety NET', which is used for the distribution of marine safety information and rescue coordination during emergencies.

Standards B and M

These are two new systems, both expected to become operational in the 1990s. Standard B will offer a similar but improved range of services to those provided by Standard A. Standard M is intended as a low cost light-weight telephone, data and facsimile service, and as such it is expected to appeal to pleasure/leisure craft operators and small vessels in general.

FINAL CAUTION

Before leaving this chapter, it is worth considering for a moment how taking any kind of radio communication equipment to sea may influence the thoughts of friends or relatives who may be concerned for your safety.

In particular, if you have been in the habit of maintaining regular contact at sea and then your friends or relatives do not hear from you for a time, it may tempt them to think (especially if they are unfamiliar with small boat sailing) that something really awful must have happened. Though this is a possibility that cannot be ruled out, it is not necessarily the most likely explanation. Electrical difficulties, interference or any one of countless other problems may prevent messages from getting through and, after all, boats function very well without radios, so they are hardly a vital piece of equipment. In short, although hearing a message can tell you a boat is safe, not hearing one does not mean it is not.

An explanation of the limitations of radio communications can save much anxiety, though of course it is impossible to say at what stage after losing contact one should initiate a search. However, this is a decision for the rescue services and anyone with any reason for suspecting that a particular vessel is in difficulties should not delay in notifying them of the facts.

3. Amateur radio

> You know I have always considered myself an amateur.
>
> Guglielmo Marconi

MARITIME MOBILE AMATEUR RADIO

Don't be misled into thinking that the word 'amateur' in amateur radio means that it is in some way second-rate. Many radio amateurs have, or have retired from, a professional interest in some aspects of the subject, perhaps as engineers or commercial operators. The number of amateurs world-wide runs into millions, and although licensing conditions are strict they have access to a wide range of frequencies stretching right across the radio spectrum. At any time of the day or night it is usually possible to hear a dozen or so, and most likely a lot more. With such large numbers of participants there is a wide spread of radio interests. Some are concerned with the pursuit of club awards, contests, emergency communications nets, low power (QRP) operation, or simply chatting to friends, but whatever activity it is, all have an equal right to their part of the amateur bands. Not all of these activities have small boat applications but an ability to keep in touch over long distances and to exchange information and ideas are uses that are easy to appreciate.

Unlike CB radio, which operates with low power on a tiny band close to the top of the HF range, amateurs are permitted to use higher powers and have access to bands in most parts of the radio spectrum. For small boats it has many benefits which more than justify the time and effort needed to obtain a licence and set up the necessary equipment, but, by its very nature, amateur radio can never substitute or even compete with professional services. The licence imposes very stringent restrictions upon the kind of messages that can be passed; any kind of business discussions are particularly forbidden. Also, unlike the professionals, amateur operators do not have to answer calls or keep regular hours. In practice they may keep regular schedules, provide much useful information and a widely appreciated service, but it has to be remembered that they are doing what they are doing simply because they enjoy it.

Amateur nets

These are radio 'meeting places' and can be a mine of information. Nets operate on particular frequencies at particular times, groups of boats calling in to report positions, enquire about the weather, marina charges, spare parts, the cost of fish or almost any other topic of mutual interest.

There are hundreds of nets operating throughout the world, many being *ad hoc* affairs that are started by small groups of amateurs who happen to be sailing together. Times and frequencies are arranged to suit themselves and the net may close down after a season or two. Others, though, are more widely known, may have been established for many years, and have active participants and listeners whose numbers are counted in hundreds.

With so many potential callers, to avoid the general mayhem of everyone calling at once, one station is appointed the key role of net controller. Its function is to allocate order to callers, and to try to arrange relay stations to help those with weaker signals that might otherwise be ignored. By no means do all net controllers operate maritime stations themselves. Many are land based and are run by individuals who freely contribute huge amounts of their time and personal resources to helping small boats at sea. Often their professionalism matches that of commercial operators but their service is not something we should take for granted. They are, after all, amateurs who do what they do as a hobby not because they are paid. It would be crass to assume that they will always be there and to make unfair demands. The success and character of nets depends very much upon the efforts of the net controllers. The fact that around the world there are several nets that have operated regularly for ten or twenty years is a tribute to their efforts.

Maritime nets may be found operating on just about any of the amateur bands, though the SSB section of the 20 metre band is the most popular. Forty and 15 metre bands are close second favourites and have the advantage that they are often less busy. There are also a few nets operating in Morse.

Exact frequencies and operating times of nets change too often for it to be possible to publish full lists of all nets, though times and frequencies of some that have become well established are given in Reference Information, section II. If you are able to transmit, once you have located one, it should be easy enough to call in and enquire about operational details of others.

The role of amateur radio in an emergency

In the kind of emergency that presents an immediate threat to a vessel and/or life, it would be totally wrong to expect amateur radio to provide the most effective means of obtaining assistance. Responsibility for co-ordinating rescue efforts is a duty performed by the coastguard or military, and though exact arrangements differ between countries they usually have direct controlling access to other services, such as lifeboats, search and rescue aircraft and other vessels in the area. Given most types of emergency, it is highly unlikely that one would find an amateur radio operator better placed to organize a rescue attempt.

Nonetheless, circumstances have occurred when amateur communications have provided the only means of effecting a rescue, though for various reasons these are not always reported in the press. One such event which did get wide press coverage occurred during the BOC solo round the world challenge race in 1983. On the night of Wednesday, 9 February, midway between New Zealand and Cape Horn, Jacques de Roux's 43 ft boat *Skoiern III* was rolled. The Frenchman was saved from going overboard by a harness but was badly bruised and bleeding. The vessel righted itself but was in no fit state to continue, though de Roux managed to raise the alarm via the Argos satellite transponder that he carried aboard. With this system the vessel's position could be located to within five miles, but the problem was that there was no shipping in the area and the nearest land was 1800 miles away. The only possible source of help was from other competitors in the race. The problem of contacting them was passed to Rhode Island amateur Rob Koziakowski. He was a disabled Vietnam war veteran who was operating from the basement of his house and who had been in touch with competitors during the preceding weeks. He, in turn, passed the request to fellow amateurs. Twelve hours later Matt Johnson, on New Zealand's South Island, managed to raise Richard Broadhead aboard *Perseverance of Medina*. He was some 300 miles beyond de Roux but none the less turned about for a rescue attempt and under the prevailing conditions at the time, the return journey would take 40 or more hours. During this time, as more water was taken aboard *Skoiern III*, the situation became more desperate; for Broadhead the job of finding him was much like looking for a needle in a haystack. However, both Koziakowski and Johnson were able to relay position information obtained from the Argos position indicators. Finally, at 19.39 on Friday the 11th, the miracle happened and the two vessels made contact. Just five hours later the Argos on *Skoiern III* stopped transmitting, indicating that the boat had sunk.

In less remote parts of the earth, where rescue services are well established, amateur radio can still have a useful part to play, but in general perhaps its best use is not as a system of last resort but as a means of preventing trouble before it happens. Emergencies do not always happen suddenly, but often as a sequence of minor events. In these cases the radio can provide a means of obtaining advice or plain reassurance before circumstances deteriorate and become life threatening.

Calling mayday or pulling the pin out of an EPIRB tells the world that you are in distress, but sometimes, though the situation is serious, it might not be an emergency. An injured crew member or a damaged rudder both have the potential to threaten the safety of the boat, though they may not place it in immediate danger. In such cases contact with another boat or one of the maritime nets could be the easiest way of getting the advice needed or may avert premature rescue or salvage attempts.

Amateur radio without a licence

No one needs a licence to listen in to amateur transmissions. It's hard to know exactly how many listeners there are on the amateur nets but on the larger ones, especially those likely to give weather information, there is evidence to suggest that non-transmitting listeners can outnumber licensed amateurs by several times. In addition to weather it is often useful to keep in touch with the positions and intentions of other boats and to know that they have reached their destinations safely. Sadly, many people only discover the value of amateur radio after they have set off on an extended cruise. Once away from your home country, taking the necessary examinations and obtaining a licence becomes very much more difficult.

Though anyone can buy amateur equipment, it must be said that other than in an emergency, without a licence, it would be totally wrong to use it to transmit. A condition of the amateur licence is that, except under clearly defined and exceptional circumstances, *radio amateurs are only permitted to pass messages to other licensed amateurs.*

From time to time, licensing authorities take action against amateurs who infringe these conditions. Those who do so risk losing their licences, fines and confiscation of their equipment. In the case of net controllers, this would be a loss to the amateur radio and marine worlds as a whole.

Obtaining a licence

The UK procedure for obtaining a licence has fairly broad similarities with that adopted by other countries. The Department of Trade and Industry publish a booklet which gives full details of the requirements (see *How to Become a Radio Amateur* in the book list starting on p. 159).

Essentially, applicants need to have passed the City & Guilds Radio Amateurs' examination. This is a straightforward examination which tests the candidate's knowledge of licensing conditions, transmitter interference, electromagnetic compatibility and operating procedures, practice and theory. A certain amount of study is required, even for those already involved in some form of electrical or electronic engineering. This can be done through local education authority evening classes, correspondence courses or, if you have the confidence, by working alone from books. No great depth of knowledge is needed and the standard should be easily attainable by most individuals – all you need is enthusiasm and persistence.

Having passed this exam, you will be qualified to apply for a class 'B' licence which permits the use of bands above 50 MHz, including VHF and UHF. These are useful for short-range communications and no doubt, as trends towards computer controlled transmissions and satellite communications continue, it can be expected to assume an increasing importance for longer distance contacts. However, most long distance communications between small boats today are carried out on high frequencies (1 to 30 MHz). To use these you need to pass a test to show that you are able to send and receive short pieces of Morse code at 12 words per minute. Newcomers often think this sounds incredibly fast but after a little experience it sounds less daunting and with practice and perseverance it is a standard well within the capabilities of almost everyone. When compared with commercial speeds of at least 20 words per minute it is slow, but it would be a mistake to believe that the skills can be acquired after a few short sessions.

Probably the most effective way of learning is in company with others at an evening class. People who attempt to teach themselves often unwittingly develop bad sending techniques. Though the code makes perfect sense to themselves it may be unintelligible to others and with more practice the habits become reinforced and more difficult to correct. To help prevent this, whatever method of learning you choose, in the early stages it is a good idea to spend a considerable amount of time listening to well-spaced Morse before attempting to send any. In this way, the correct character, letter and word spacings become instinctive and you develop a natural tendency to send in the same style.

Though commercial maritime stations are slowly giving up the use of Morse, just about all countries which issue an amateur licence only allow those that are proficient in Morse to have access to HF bands. Any significant relaxation of this condition seems unlikely. This is far from being a boring piece of bureaucracy, as the code has special advantages for those at sea:

1. Only the simplest of radio equipment is needed.
2. Very low power levels may be used to communicate over long distances.
3. When atmospheric conditions are poor, or when bands are close packed with other stations, Morse succeeds when most other communication modes fail.
4. The use of internationally recognized abbreviations (see Reference Information, section I) makes it possible to communicate across language barriers.

Rather like swimming or riding a bicycle, Morse is one of those skills which once learnt is never forgotten. It appears to be retained at some subliminal level, and though the practice required to work up an effective speed can be time consuming, it is time well spent.

NOVICE CLASS A AND B LICENCES

These have been introduced to encourage young people to take up radio as a hobby. Examination requirements are less demanding, but attendance at a recognised practical course is a pre-requisite. Both licences give access to a limited range of amateur frequencies and in the case of the A Novice licence this includes some from the HF bands. In this case (like the full A licence) a morse test pass is a further requirement though the speed is only 5 words per minute.

Other restrictions include a limit of 3 watts on the transmitter output power and the exclusion of maritime mobile operation. However, licence holders may operate mobile stations from vessels on inland waterways.

CALL SIGNS

Call signs are intended to help identify sources of radio interference and act as an aid to licence administration. They are issued to new amateurs along with their first licence. Although there are circumstances under which they can be changed, this is most unusual, and in most cases they remain with the same operator indefinitely. Each call sign is unique but in general they are made up as follows:

1 or 2 characters + a single digit + up to 3 characters

The first characters are used to identify the country of issue while the characters that follow may refer to a particular region within the country of issue, the class of licence and its date of issue.

MARITIME MOBILE OPERATION

Most amateurs operate shore-based stations but many countries issue a licence which also covers maritime use. This permits use at sea, in international waters and within the territorial limits of the country of issue. In these cases the use is identified by stations adding the suffix /MM to their call sign.

Reciprocal licences

In other countries, and within their territorial limits, it is usually necessary to obtain a reciprocal licence, but arrangements for this differ considerably between countries. In some cases, if reciprocals are possible at all, administrative procedures are tedious, time consuming, and may be expensive, though things are not always this difficult. In the case of certain European countries, arrangements are particularly straightforward (see CEPT licences below) and in others an application may simply involve giving details of your existing licence to the Telecommunication Administration of the country you intend to visit. However, contact with the National Radio Society of the country concerned is probably the easiest way of finding out exactly what is required.

CEPT licences

Under a European arrangement, a growing number of countries issue CEPT licences (European Conference of Postal and Telecommunications Administration). These enable amateurs to operate portable or mobile stations within countries that have implemented the CEPT Recommendation TR/ 61–01. The number of these countries is growing rapidly and at the time of writing includes:

Austria	Belgium	Switzerland	Germany
Spain	France	Liechtenstein	Luxembourg
Monaco	Norway	Netherlands	Sweden
Denmark	Finland	Greece	

Amateurs taking advantage of these arrangements are required to operate within the terms and limitations in force both within their home country and in the country that they are visiting.

Future amateur radio developments

With the application of computer technology, the field of amateur radio is expanding. At present most maritime nets are held as SSB voice contacts, and no doubt this will remain popular. However, as marine users become aware that the ability to send and receive computer data can greatly increase their access to useful information so more use will be made of these data modes.

4. Data modes

On tuning across HF bands, many of the transmissions that you hear will be in code. Sometimes they are mistaken for interference rather than any kind of sensible broadcast but they hold a wealth of information for the small boat user. Weather reports, fax pictures, agency news and navigational information are examples but different types of codes are used for different kinds of data transmission. Codes used for NAVTEX differ from those used for fax. Many maritime radio stations still transmit weather forecasts in Morse, and various forms of RTTY are used by other stations reporting news and weather.

To the casual listener these data transmissions may simply sound like interference as, apart from Morse, they are intended to be decoded by machine. Even so, each has its own particular sound and with a little experience it is quite easy to identify some of the different types. Fax is a bit like someone scratching a knife against a brick, whilst packet radio makes an intermittent *brruuuuuuuup* noise. ARQ AMTOR has a repetitive chirping sound and RTTY has a sort of musical sound, not unlike that made by a few stones behind the hub cap of a car driven at speed. These sounds are all very difficult to describe in words but so obvious once you have learned to identify them.

Some data transmissions are not intended for public reception. These may contain sensitive military or commercial information and are encrypted to conceal the information they contain. Apart from these special cases most data transmissions can be turned into an intelligible form and, using modern technology, the equipment required can be quite small and need not consume large amounts of power.

EQUIPMENT FOR DECODING DATA TRANSMISSIONS

In the past, the only way of working with data transmissions was to buy a separate piece of equipment for each mode, i.e. a NAVTEX machine for NAVTEX, a telex machine for telex, a WEFAX machine for WEFAX, etc. On vessels with sufficient space and power to accommodate the equipment this approach does have certain advantages:

1. Failure of one item does not affect the others.
2. Operating procedures can be kept simple.

These days there is an alternative approach made possible by microelectronics and by the fact that, in spite of superficial differences, these machines share several functional similarities. All need a radio to receive or transmit the data, a screen or printer for displaying received texts and a keyboard for typing in text or commands. A portable computer can provide the display screen and keyboard, but more than this, with appropriate software it can carry out the decoding needed to convert raw data signals from the radio into fax pictures, NAVTEX information or telex messages, etc.

Most computers are fitted with a serial port for connecting peripheral devices to other computers but very few have any facility for direct connection to a radio set. To overcome this an interface device is needed to connect between the two, and various types are available, but before considering these in more detail, let us first look at some desirable features to have when choosing radios and computers for these modes.

Radio requirements

Most good quality receivers (or transceivers) can be used for receiving data transmissions, but they should be capable of being tuned to within 1 kHz. Within 0.1 kHz would be better and the ideal would be one tenth of this. Once set to a particular frequency it is important that the receiver does not drift as this will make tuning of some transmissions frustrating, if not impossible.

If you are using a transceiver and intend sending data an important point to remember is that the transmitter will be expected to work much harder than when sending SSB speech.

More current will be drawn from the power supply and extra care must be taken to ensure a good match between the transmitter and antenna. (See Chapter 7, p. 111.)

In using packet radio, ARQ AMTOR and related codes (see p. 53), another important consideration is the transceiver transmit to receive (and receive to transmit) switching times. (These, incidentally, may not be equal.) With AMTOR the transition must be accomplished in less than 50 ms, but with commercial semi-duplex versions of the code there is a further consideration. Here the transceiver has also to change frequency as it switches between receive and transmit and, unfortunately, many (particularly older) transceivers are incapable of doing this quickly enough. Sometimes the speed can be achieved by making modifications to the set and in these cases manufacturers or distributors may be able to advise. This, though, is a point to check if you are buying a transceiver that may not be specifically intended for working these modes.

Computer requirements

Computer requirements are not particularly demanding. Of first importance is the serial data port (usually type RS-232) already mentioned but, this apart, other desirable features are low power consumption, robust construction, freedom from electrical noise and a good clear, clean display. Colour is by no means essential.

A printer for making hard copies of received texts can be a useful extra but for most purposes a disk drive is sufficient. Storing text or fax pictures on disk gives a considerable saving on space when compared with paper storage but is slightly less convenient as the computer has to be used to redisplay the data.

COMPUTER-GENERATED RADIO INTERFERENCE

A problem, particularly with some early computers, was that they produced so much electrical noise that listening to a radio in the same room was difficult. When the two were connected together the racket became so loud that any kind of reception was impossible. (See Chapter 6, p. 95.) Some BBC computers appeared to be particularly prone, producing much broad band interference. Fortunately, manufacturers are now more aware of the difficulties and the most recent portable computers present less of a problem. Even so, there can be difficulties with spot frequencies.

Before buying a computer for radio use it is as well to seek the advice of others already using one to see how they have coped with any interference difficulties that may have occurred.

THE RADIO/COMPUTER INTERFACE

The purpose of the interface is to convert audio signals from the radio to a digital form that can be accepted by the computer. They are variously referred to as decoders or demodulators, and those that also convert digital signals from the computer into signals for transmission may be known as modems, terminal node controllers (TNCs) or data controllers, depending on their application.

Decoders for reception only

If you require nothing more than to be able to receive, say WEFAX pictures, one of the simplest means of achieving this would be to use a software program that would enable the computer to carry out the decoding functions. An example of this type of program is PC-HF-FAX by John Hoot which fits easily on a 5.25 in or 2.5 in disk. This enables FAX pictures to be displayed on screen, stored on disk or dumped to a printer. Sections of pictures can be enlarged and it also has a facility to aid tuning. Supplied with the software is a cable for connecting the radio extension speaker output to the computer serial port and this includes a small number of components needed to convert the audio signal to a form acceptable by the computer. This program is written specifically for IBM compatible computers and an alternative version is available for decoding Morse, RTTY, NAVTEX and AMTOR.

Photo 4.1 PK–232 multi-mode data controller. (*Courtesy of ICS Electronics Ltd*)

Multi-mode data controllers – for reception and transmission

These devices are microcomputers in their own right and dedicated to the purpose of transfering data between radio and computer. When compared with the software decoders described above, multimode controllers are considerably more complex to set up but have the advantages of being able to operate in conjunction with many different types of computer and offer very comprehensive capabilities. One example is the AEA PK-232 shown in Photo 4.1 which is capable of processing Morse, ASCII, AMTOR, SITOR, NAVTEX, RTTY, FAX and packet radio codes.

Because these units are intended for general markets, rather than marine users in particular, their design is extremely versatile. Describing their full capabilities and the variety of ways in which they may be configured would take many chapters but for small boat users, their main attractions are:

1. Virtually all data modes can be decoded with one piece of equipment.
2. Space and electrical power requirements are low.
3. A single antenna can be used for all data modes. Not only does this avoid the forest of antennas and their associated wiring frequently found aboard boats, but thieves are not provided with a 'list' of the kind of equipment they are likely to find aboard.
4. The computer and transceiver or general coverage receiver that form part of the installation can be used for other purposes in addition to decoding data.

CONNECTING A DATA CONTROLLER TO A RADIO AND COMPUTER

This is usually a straightforward job as there is a basic similarity in the connections needed for most types of data controller, computer and radio, though there are many differences in plug and socket requirements to pin configurations. In many cases an audio output signal from the radio's extension speaker socket or phone jack will provide a suitable input signal for the decoder.

If you are using a transceiver and intend transmitting data, two additional connections are needed. These are the audio signal connection from the decoder, along with a connection to the push

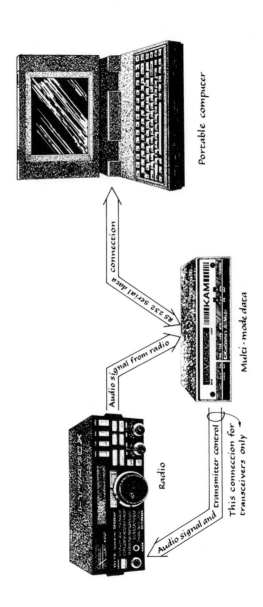

Fig. 4.1 Interfacing a data controller to a radio and portable computer.

to talk switch which is used to control the transmitter. In most cases both of these extra connections can be made via the transceiver microphone socket.

Connecting the serial interface is largely a question of making sure that you join pins on the data controller interface to the corresponding pins on the computer. Screened cable should be used for the interface connections, which in most cases will be terminated in a 25-way plug and socket. A possible difficulty here is that not all manufacturers have adopted the same pin configuration for their interface connections so these details must be checked with both the computer and data controller manuals. Of the 25 pins used in the interface probably only about six will be needed but the exact number may depend upon the software that you have chosen to use with the controller.

SOFTWARE FOR DATA CONTROLLERS

A software communications program is required to enable the computer to communicate with the controller through its serial data port. Its purpose is to:

1. Send command codes to the decoder. These are used to set up various aspects of its performance, such as the mode in which it is to work, the speed at which it is to operate, etc. (Some specific commands will be mentioned in more detail later.)
2. Show how the existing command states of the controller have been set.
3. Display on the screen messages that have been decoded and any waiting to be sent.

Such software need not be particularly complex and is sometimes written in Basic, though when so much is available in the public domain, it is hardly worth writing your own. Communications programs intended for sending electronic mail on telephone systems can often be pressed into service.

Newcomers to the field can sometimes save themselves a great deal of trouble by ordering their communications software directly from the data controller suppliers. By specifying the kind of computer with which it is to be used, you may expect to obtain a more user-friendly set of software and save the frustration that may occur when trying to adapt a general purpose communications program.

Data modes

Photo 4.2 Worldwide communcations. In addition to the normal functions of this HF transceiver (top), with the multi mode data controller (located below the transceiver) and a portable computer (bottom) it can function as an automatic telex terminal, can transceive packet radio, ASCII, RTTY, Morse or AMTOR codes and receive NAVTEX or WEFAX pictures.

Note that software that will handle Morse, RTTY, AMTOR and packet radio may not necessarily have the graphics capability required to display fax pictures. If you wish to receive WEFAX pictures, do make sure that this facility is included. A reasonable fax program should have the capacity to:

- Display pictures on the screen as they are being received.
- Store them on disk either whilst they are being received or when reception has been completed.
- Print displayed or stored pictures.
- Clear the screen or picture storage at any time.
- Enlarge or reduce parts of pictures.

Data decoders and controllers in use

At the time of writing there are three prominent manufacturers of multi-mode decoders – AEA, Kantronics and MJF. All share many functional and operational similarities. In the remainder of this chapter we take a general look at how they are used, but where particular examples are given, these refer to the Kantronics KAM multimode. This apart, very much of what follows is about data modes in general and so is not tied to any particular kind of equipment. With such a wide subject the object of this section is simply to provide an introduction to what is possible rather than detailed instruction on how it is achieved.

PRELIMINARY SETTING UP

Once you have the radio, computer and decoder connected up, the first job is to program the serial port on the computer so that data can be exchanged between the computer and the controller. This involves matching the data format provided by the controller with that of the computer. This may be achieved through software commands or by setting switches on the computer or controller. The exact procedures vary and will need to be checked with the computer, software and controller manuals. In any event it will not be possible to proceed until this has been done. In general there are four variables that need to be set and these are as follows:

        1. Baud rate*   300, 600, 1200, 1800, 2400, 4800 or 9600
                           (Use 1200 or 9600 for fax reception)
        2. Parity   None

3. Data bits   8
4. Stop bits   1

* **NB** The baud rate mentioned above is that used to communicate between the computer and decoder. This may well be different from the baud rates used to transmit or receive RTTY or fax, etc. These speeds are set up later and involve separate commands. See below, under the sections for each particular data mode.

When the link between the computer and controller has been established, the next step is to get the controller in the command mode ready for receiving the command instructions which set its internal parameters. Typically, this condition is shown when the characters *cmd:* appear on the computer screen, but the procedure for achieving this may vary between controllers so you will need to check with the manual if the prompt does not appear immediately.

COMMAND CODES

Command codes are English-like words that form abbreviations for particular aspects of controller behaviour. In the Kantronics KAM, for example, typing the WEFAX code prepares it for receiving weather fax pictures. Similarly, the code RTTY sets things up for telex, and CW for receiving or transmitting Morse. In total there are some 175 command codes each affecting separate parameters. When setting up a controller for the first time, be prepared to spend time studying these codes as successful results will very much depend upon setting them up correctly. Once this is done, many remain stored in the controller memory for subsequent use and rarely need to be reused. Usually this part of the installation is the most time consuming but fortunately, in practice, most users find that very few of the codes need be committed to memory; on the odd occasions that others are needed they can always be looked up in the manual.

Broadly, command codes fall into two groups. Those, such as the three mentioned above, which make an immediate but temporary change to the controller's performance and those whose effects are more long term.

Some examples are:

ABAUD	This parameter is used to set the baud rate used for data transfer with the computer, via its serial port.
MYCALL	Used to store the operator's call sign, which in some modes may be transmitted automatically.
INVERT	Reverses the mark/space signals on received RTTY, AMTOR or ASCII codes (polarity reversal).
FSKINV	Reverses the mark/space signals when transmitting RTTY, AMTOR or ASCII codes.
ECHO	When turned on, this command causes everything that is typed on the computer to be displayed on the screen. This can be useful if your communications program does not provide a split-screen display on the computer. In these cases, one part is dedicated to displaying instructions typed in on the computer, whilst the other is reserved for messages received from the controller.
MYSELCAL	This command is used to enter a selective calling code; which, when transmitted by another station, will cause your equipment to respond by leaving its standby condition and become ready to receive the call (used in AMTOR and SITOR).

The codes are used with the format **CODE XXXXX** where XXXXX is a new value to be assigned to the code. However, if this part is omitted, the computer will respond by showing the way in which the code has already been set. Typing ECHO or MYCALL, for example, would produce a response on the screen which would tell you the current settings, as follows:

Commands typed in	**Screen display**	**Comment**
ECHO	cmd:ECHO OFF	Or ON, as the case may be.
MYCALL	cmd:MYCALL G0HOC/G0HOC	Indicating the call G0HOC.

To change the settings you simply type the code, followed by the new value, and the response would be as follows:

ECHO ON	cmd:ECHO was OFF	Or ON, if it were already on.
MYCALL ZB2IS	cmd:	Only the command prompt is displayed but internally the call sign has been changed.
MYCALL	cmd:MYCALL ZB2IS/ZB2IS	

These and many other codes can remain stored within the decoder only as long as the power supply remains connected. Turning off the power supply will cause them to resume their former values the next time the unit is used, but there is a command, PERM, which is used to place them in longer-term storage. Whenever this command is used, any long-term variables that have been changed are stored in EEPROM (electrically erasable programmable read only memory), ready for the next time the unit is used.

DECODING – GETTING STARTED

Morse

Because Morse transmissions are easy to recognize, they are perhaps a good point to begin exploring data modes. You might begin with a weather forecast from a commercial station, such as Portishead Radio (see p. 147), or try some amateur Morse which is found at the bottom of most bands (e.g. 14.000 to 14.070 MHz) but first there are some command code settings that need to be checked. These are:

CWSPEED	10	Sets the Morse speed at which the decoder will operate when first put into the CW mode.
CWBAND	200	Sets the audio frequency bandwidth of the controller input filter. Reducing the value makes signals harder to tune but gives more reliable copy. (Values between 10 and 1000 are acceptable).
CWTONE	750	Sets the centre frequency (in Hz) at which Morse will be received (other values between 50 and 500 can be used).

Typing the command CW, at the cmd: prompt places the unit in the CW mode and, once a morse signal is tuned correctly, the decoded message should begin to appear on the screen. If you do not achieve success the first time, do not despair. One possibility is that the receive speed does not match that of the incoming signal. In this case the receive speed can be unlocked and the controller allowed to track the speed of the incoming signal. This is done by holding down the control key and pressing C, then releasing both keys and pressing U (Ctrl-C U). Other points to check are that the signal is strong enough and tuned correctly. Failing this, you might try increasing the value of CWBAND.

Do be prepared for some trial and error as, with all data modes, a little practice is needed to obtain good results. Once you have discovered and PERMed the command codes that best suit your own equipment and personal preferences, decoding Morse becomes a routine process and takes only a few moments to set up.

When reading Morse texts, you are likely to find many abbreviations. These may not occur to a great extent in weather forecasts, or other broadcasts that follow a fixed format, but in other transmissions operators often make extensive use of the international Q code. This not only speeds up communications but also helps overcome some of the difficulties of language barriers.

(See Reference Information, section I for a selection of Q codes and other abbreviations.)

Decoding RTTY and ASCII

RTTY is the radio equivalent of wire telegraphy (see p. 4, Chapter 1) and is used extensively for weather and news reporting. It uses the same five-bit code, transmitted by frequency shift keying. Tuning in to RTTY signals is a critical job, made easier by the tuning indicator provided on most data controllers. Decoding is somewhat complicated by the fact that, throughout the world, various stations have chosen to transmit at different speeds (bauds rate), to use different frequency shifts and occasionally to reverse (i.e. invert) the polarity of mark and space bit frequencies.

The following table shows the relationship between speed and words per minute:

Bauds	Words per Minute
45	60
50	67
57	75
75	100
100	132
110	147
150	200
200	267
300	400

One or two data controllers are capable of automatically self-adjusting their parameters to match the characteristics of the signals to which they are tuned. For controllers without this facility it is a question of sequentially working through different combinations of speed, shift and polarity, until the screen display becomes meaningful. This involves a certain amount of guesswork and trial and error, but a few starting points are given in the following table.

Application	Baud	Shift (Hz)	Polarity
Amateur	50	170	normal
News services	50	170 or 425	various
Aeronautical, military and weather	various	850	various

After trying all possible combinations of shifts, speeds and polarities, there will still be some transmissions that defy all efforts at decoding. Apart from those that are deliberately encrypted to preserve secrecy, transmissions from Soviet, Eastern Block and Arab countries may use codes containing Cyrillic or Arabic characters.

ASCII

This seven-bit code is used internally by many computers and can encode upper and lower case letters, along with punctuation and a selection of symbols. In the form in which it is transmitted, an eighth bit is added, which is used for parity checking, so giving some possibility for detecting errors. Most users are amateurs who treat it as an alternative to ITA no. 2 but, as it is incompatible with older electro-mechanical terminals, it has not gained general popularity.

TOR Codes – AMTOR, SITOR and NAVTEX

TOR codes are used extensively by commercial radio stations for broadcasting traffic lists, weather information, NAVTEX and in exchanging telex messages between ships and international telex networks. These codes are especially good at getting through poor radio conditions and part of the reason for this is that they contain features which help correct any misunderstood (error) characters that the receiving station may pick up.

Unlike the ITA no. 2 code which uses 5 bits, these codes use 7 and every character is expressed as some combination of 3-space and 4-mark bits. The receiving station checks the numbers of mark and space bits and any discrepancy is interpreted as an error. The usual transmission characteristics are 100 bauds and 170 Hz shift.

Two kinds of TOR versions are in regular use and these are referred to as ARQ (Automated Repeat Request) or mode A and FEC (Forward Error Correction) or mode B. Which type is used will depend upon on whether the transmission is intended for reception by a single station (e.g. telex messages) or for reception by many stations (e.g. NAVTEX).

ARQ

Once you have heard the repetitive chirping sound made by stations using this mode, it's hard to mistake it for anything else. The way that the system works is that the information sending station transmits a block of three characters. When these are picked up by the receiving station, they are checked for errors and, if none are detected, a control code is transmitted back to the sending station, which then transmits another block of three. If an error is detected, the receiving station transmits a different control code which causes the sender to repeat the previous group. In this way, both sender and receiver transmit repetitively, hence the characteristic ARQ chirp.

At times when other kinds of radio communication are blighted by fading, interference or static, the ARQ mode can still be effective. Such conditions may lead to many receiving errors and repeat requests, so communications will be slower, but error-free copy may still be achieved.

FEC

The sound made by FEC transmissions is somewhat similar to RTTY. Because it is intended for reception by many stations it is not possible for receiving stations to make automated repeat requests if they did not get it right first time. Nonetheless, a certain amount of error correction is still possible. The way it works is as follows:

Two streams of characters are taken from the text but one lags five characters behind the other. The sender transmits alternately from these two streams and in this way all characters are sent twice. As with ARQ, the receiving station checks for errors by counting the numbers of mark and space bits in each character and rejects those that do not conform. If neither of the paired characters conform, the computer indicates this by placing a missed character indicator in the screen text.

AMTOR *(Amateur microprocessor telex on radio)*

AMTOR was developed from the marine TOR codes in 1979 by Peter Martinez and is today widely used by amateurs who, unlike commercial marine users, usually transmit and receive on the same frequency. As well as supporting the usual ARQ and FEC modes, the system also supports a third mode, LAMTOR (or mode L), which is used for receiving only. Its purpose is to provide a means of listening in to ARQ and FEC communications but in the case of ARQ no repeat requests are possible. Consequently, any characters missed on LAMTOR must remain lost. Furthermore, any repeats made by communicating stations will be displayed, so the received copy sometimes looks a little strange. Nonetheless, it may still be intelligible, if viewed with these points in mind.

NAVTEX *(transmitted on 518 KHz)*

This international service is sponsored by the International Maritime Organization (IMO) and provides mariners with meteorological information. Information is collected from various sources by a NAVTEX coordinator. These are then sent, usually by telex, to appropriately located NAVTEX stations for transmission on 518 KHz. This frequency is used by all NAVTEX stations but they avoid interference by using different time schedules. (See Reference Information, section II, for details.)

Though the service is not currently available everywhere in the world, it is hoped that eventually it will be. Areas covered include:

The Baltic and North Seas
Mediterranean, Aegean and Black Seas
USA East and West coasts
Atlantic coasts of Spain and Portugal
Azores
China
Japan
South America

MESSAGE FORMAT

Messages are transmitted in English and follow a set format. First, there is a preamble consisting of four character groups, B1, B2, B3 and B4 in automatic receiving equipment which are used as a basis for sorting messages. These are coded as follows:

B1 – Transmitter identification
   This is a single letter identifying the station,
   e.g. F = Brest, L = Scheveningen, S = Niton
B2 – Subject identification
   Another single letter for identification:

A	Coastal navigational information
B	Gale warning
C	Ice report
D	Distress (search and rescue) information
E	Meteoroligical forcasts
F	Pilot service message
G	Decca system information
H	Loran C system information
I	Omega system information
J	Satnav system information
K	Other electronic nav. aid messages
L	NAVAREA warning
M to Y	For service use or not assigned
Z	No messages on hand

B3 and B4 – Message numbering
These two characters form numbers within the range 01 to 99 and are used to identify particular messages within a subject group. After 99, numbering begins again at 01 and, to avoid the need to repeat any still in force, users are expected to have facilities for storing messages.

Next follows the message text and the following example shows the kind of format used:

```
ZCZC GB31
CULLERCOATES
GALE WARNING THURSDAY 8 FEBRUARY
2250GMT
NORTH UTSIRE
SEVERE GALE FORCE 9 BACKING
SOUTHWESTERLY SOON

SOUTH UTSIRE FORTH TYNE DOGGER
FISHER
SEVERE GALE FORCE 9 NOW DECREASED GALE
FORCE 8 BACKING SOUTHWESTERLY
SOON

GERMAN BIGHT
WESTERLY SEVERE GALE FORCE 9 DECREASING
GALE FORCE 8 IMMINENT

THAMES SOLE
GALES NOW CEASED

NNNN
```

DECODING

The transmission mode for NAVTEX is FEC SITOR, and so decoding with AMTOR or LAMTOR should not present difficulties. However, when using decoder modes not specifically intended for NAVTEX, the preamble codes B1 to B4 will be ignored and so every message received will be displayed. There may be a considerable number of messages and though this may present no difficulty to those using disk storage, if printed paper is the only available

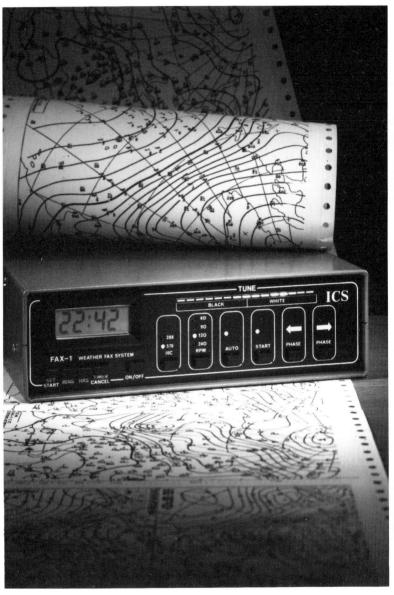

Photo 4.3 The ICS FAX–1 decoder. This decoder contains all the necessary microelectronics and software for decoding NAVTEX and fax. A radio receiver is required but no additional computer is needed as the decoder output can be sent directly to a printer. (*Courtesy of ICS Electronics Ltd*)

storage medium the amount needed may be unacceptably large. In any case, many messages may be of no interest. To avoid these problems some decoders are fitted with NAVTEX software, which includes a facility that allows certain types of message to be ignored. Navigational information, meteorological warnings and search and rescue messages cannot be ignored, but users may elect to reject other types if they feel they will be of no interest.

Weather fax (WEFAX)

Many stations throughout the world transmit meteorological data on HF (and sometimes LF) as fax pictures. The information they provide may cover a particular group of users, e.g. aircraft, military, etc. The type of information includes:

> Surface wind speeds
> Wind speeds at various altitudes
> Surface pressures
> 12-, 24-, 48- and 72-hour surface prognoses
> Ice reports
> Sea swell forecasts
> Tropical cyclone warnings
> Mean sea surface temperatures

Some stations transmit redigitized satellite pictures but many weather fax pictures are hand-drawn charts prepared from information collected from land-based stations, satellites or other sources. In original form, full-size charts are 15 in wide and between 10 and 18 in in length, but in transmitting these over HF some resolution is lost.

In past years encoding for radio was carried out on a rotating drum which was scanned at 60 scan lines per minute. Today encoding is carried out digitally on a flat bed scanner, which scans at a rate of 120 lines per minute. This gives a resolution of 96 lines or pixels (picture elements) per inch, resulting in a maximum number of 1800 × 1800, or 3,240,000, pixels. As this is far more than could be displayed on a portable computer screen, one task for the decoder software is to reduce this to a level that can be displayed without too serious a loss of detail.

TUNING WEFAX SIGNALS

In tuning WEFAX signals, set the receiver to upper side band and a frequency 1.7 kHz below the published frequency.

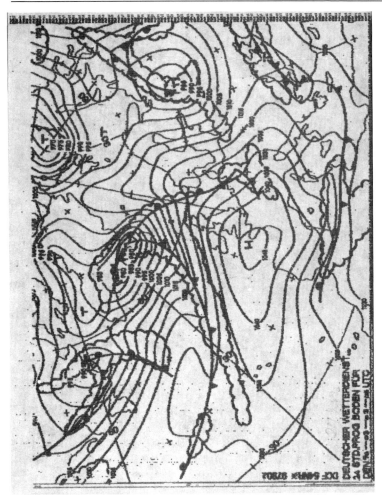

Fig. 4.4 A sample print-out of a weather fax picture received on amateur equipment.

Section III of the Reference Information gives frequencies and times of operation of some well-known fax stations. Most transmit pictures on a repeating 24-hour schedule and during this period some stations will include a transmission giving details of their schedule (i.e. which pictures to expect at what times).

Amateur packet radio

Packet radio is a new data mode and is still very much in an evolutionary stage. Unhampered by needs to maintain compatibility with old electromechanical equipment, it makes effective use of microelectronics and computer technology to provide an error-correcting system capable of sending the full ASCII character set. In addition, computer power is used to carry out a range of automatic management facilities of which the operator is largely unaware. This, though, is only a small part of the story and in a brief discourse such as this it is possible to give only a brief outline of its potential.

The term 'packet' is used because message texts are sent, not in a steady stream, as with CW or RTTY, but in blocks or packets. A typical packet may be several words or even a sentence or so but, before it is transmitted, other data is added. This contains information such as the identities of sending and receiving stations, and a code that enables the receiving station to check for errors. In the event of any being detected, the receiving station transmits a repeat request which causes the sending station to retransmit the packet.

PACKET FEATURES

One feature of packet radio is that even though sending and receiving takes place on one frequency, for users it feels like full duplex, i.e. either one of a pair of people in contact with each other can type messages at any time. Furthermore it is possible to hold several independent conversations on the same frequency without interference. How can this be? The explanation is that with packet radio, text can be sent more quickly than anyone can type messages, so for most of the time the computer and radio set spend their time simply waiting for the next packet.

To enable several stations to work the same frequency at the same time, the system has several facilities for avoiding interference. Each station monitors the frequency and will not send data whilst it can hear another transmission in progress but there is the possibility that collisions will occur if two stations choose exactly the same moment to transmit. In this case both stations wait a random length of time before retransmitting.

Another way in which spare computer/radio time is utilized is through the technique of digipeating (a contraction of digital repeating). By this method operators are able to use intermediate stations, which may have little idea that the process is taking place, to relay their messages to target stations well beyond their normal working range. At first this could sound like an intrusion, but across the world amateurs have established thousands of such digipeaters whose function is simply to relay messages.

Although packet radio is not the best mode for holding one-to-one conversation, it is ideally suited for passing electronic mail. Many of these relay stations operate 24 hours a day and are set up with large memory computer systems able to act as mail boxes, or forward correspondence, at times when frequencies are less busy. Through these, users have access to bulletin boards and can send or store large blocks of text or computer software for others to pick up at their convenience.

At present packet radio is not greatly used by maritime amateurs but the mode does have many advantages for the kind of traffic currently passed on marine nets. Remotely updated bulletin boards could be used to store weather information ready to be picked up when needed. Other notices could include items of navigational or general interest left by users, and individuals could contact friends by leaving messages with times and frequencies on which they will be listening.

5. Radio equipment

From time to time one hears of slightly odd people who are able to pick up radio signals through a bed frame or through the fillings in their teeth! Unfortunately, for the rest of us to enjoy the benefits of radio, a certain amount of equipment is required, which is the subject of this chapter.

In recent years the market for radio and electronic equipment has expanded enormously. Complexity has increased, equipment has become cheaper, and it would take a far larger book than this to describe fully the variety that is currently available. To restrict the field, this chapter is concerned with only the most basic types of general purpose receivers and communications equipment. The aim is to explain in basic terms some of the essential features to look for when buying equipment, rather than to reproduce the kind of material best provided by manufacturers' and chandlers' catalogues.

The chapter ends with a short section on home-constructed equipment. Someone once said that if you could not fix a piece of equipment yourself, it should not be on board. Though many people might not take quite such an unqualified view, I would not care to criticize the sentiment. As far as radio equipment is concerned there are possibilities for any one of average practical abilities to build a certain amount themselves, and this section introduces just a few.

PORTABLE RECEIVERS

By a portable receiver, I mean the kind of battery-operated set used mainly for listening to broadcast stations, weather forecasts and general entertainment, the kind sold by high street stores throughout the world. On the whole, they are manufactured for wide markets and can offer very good value for money when compared with anything made specifically for the marine market.

The problem in choosing a good one is that, without the facilities of a well-equipped test laboratory, it is practically impossible to compare the technical merits of one with another. Even when manufacturers go to the trouble of providing a specification, objective comparison may still be impossible as they often use different ways of expressing the same aspects of performance. Furthermore, once in the dealer's showroom you can expect little real help from sales staff. More often than not they are more geared to promoting their goods in terms of the fashionable styling of the case rather than its technical assets and they may have difficulties in appreciating the requirements of mariners.

As with the purchase of any complex item, the best way to avoid buying a lemon is to make sure you are well informed beforehand. Reading through the reviews of latest models is helpful, as is contact with other people who have similar radio requirements. But what exactly are the features one should look for in a radio for use at sea?

How is it going to stand up to life at sea?

Knocks, bangs and the occasional dowsing with salt water are all part of life at sea and, without protection, most portable radios will not stand this treatment for long. Some, though, are better prepared than others and sets with thin flimsy cases and long projecting knobs are obviously more vulnerable. Most sets are likely to need some extra protection against physical damage, especially if they are in an open boat or cockpit. Photo 5.5 shows one method of achieving this – the plywood box and foam rubber liner give extra protection to the Sony receiver.

Unless you have a very dry boat and the set can be given a permanent home below, protection from sea water is essential. An easy and effective way of providing total weather protection is by placing it in a heavy duty plastic bag, and here the 'Aqua Sac', or some similar type, is most suitable. These are sold by chandlers in various shapes and sizes and the types made for documents or first aid kits are often large enough to enclose most portable radios. On many sets, especially those with push-button controls, and where a telescopic antenna is not required, it is possible to use the set through the bag and in this way it need only be removed for battery changes.

Different types of fastening are used to seal the bag and these have various degrees of permeability to sea water. In the most effective kind a fold in the bag is clamped between two strips of hard plastic. It is claimed that this will exclude water even after fairly long periods of total immersion, though it is quite probable that buoyancy added to the radio by air trapped within the bag would cause it to float.

Battery life

Battery life expectations vary considerably, depending on the type of radio you have. Some consume more current than others but, apart from this, the life of a set of batteries will depend upon the size of batteries used. In the worst cases you might expect only 6 or 8 hours of life, but this may be acceptable if they are small or few in number. Rechargeable batteries can be an economic alternative to conventional disposable cells, but because of their lower terminal voltage they may not be successful in some radios (see p. 83).

With some designs, using a lower volume will consume less power; intermittent rather than long continuous use also helps to increase the running time, though any use of dial lights will have the opposite effect. On the whole, portable radios use little power and if they can be run from the boat's main power supply, running costs may be negligible. However, if the boat's supply is a nominal 12 volts and the radio requires something less, a voltage adaptor (regulator) will be needed to make the conversion. (See the section on voltage compatibility, p. 85.)

Single side band reception

As the short-wave frequencies become more crowded with stations, the trend towards SSB is going to continue, and public broadcasts are planned for the 1990s. However, to make them intelligible the set must be fitted with a Beat Frequency Oscillator and/or Upper Lower Side Band Switch. For marine use it is an especially important facility as this is the mode used by commercial stations for broadcasting weather and other information as well as amateurs running maritime HF nets.

To obtain reasonable reception of SSB signals, fine tuning is also essential. It should be possible to resolve frequencies within 0.5 KHz and it's not a bit of good if the slightest touch on the knob causes the tuning to jump by several KHz. On conventional analogue tuned sets (i.e. the kind with a knob and pointer that sweeps across a scale) this requires quite a high standard of mechanical engineering, beyond that usually found on portable sets. Reception of AM broadcast stations is not nearly as demanding and a close look at the frequency spacing on the tuning scale will give an idea of what applications the manufacturers envisaged.

Frequency coverage

If you intend to travel widely, a good range of frequencies will certainly be useful. Continuous coverage of 0.15 to 30 MHz on AM, and 76 to 108 MHz on FM, is fairly comprehensive and should cover most services but, particularly on older analogue-tuned sets, this is generally split into several separate bands. Typically, FM, LW, MW and a number of short-wave bands may be included but look for the gaps between them. If the radio is aimed at the domestic market, the chances are that many commercial marine and amateur frequencies may be excluded. Also, beware of sets that claim to include a 'marine band'. Marine frequencies are spread throughout the HF spectrum and the chances are that these particular sets will only cover those frequencies around 2.182 MHz – the MF distress and calling frequency.

In recent years an increasing number of sets have appeared on the market with digital tuning. These are radically different from traditional analogue-tuned sets and promise higher standards of performance. Although the technology is still improving, some digital sets do not always compare favourably with their lower priced analogue counterparts.

One advantage claimed for digital sets is that they are convenient. They have a key pad and if you know the exact frequency of the station you want to listen to, or if you want to quickly change to another, you just punch in the frequency and up it comes. Sometimes there is also a knob that can be used to search for stations by scanning through frequencies and this functions rather like the tuning knob of an analogue set but differs in that the frequency is not changed smoothly but in a series of small steps. In the better quality sets these steps may be so small as to be almost indiscernible (15.6 Hz in the Lowe HF 125) though in others they might be as large as 5 or 10 KHz. This could be alright if you only need to tune to stations whose frequencies divide by 5 or 10 KHz. Sadly, most are not so conveniently located, and to cover these intermediates an extra knob is usually provided for finer tuning. However, on some of the poorer quality sets this still does not give sufficient resolution for tuning SSB signals.

Frequency stability

Frequency stability has already been touched upon in the last chapter but it is also important in SSB receivers. The electrical characteristics of most pieces of electronics change with temperature. Radio sets are no exception and designers try to ensure that their sets are not seriously affected by temperature changes within the normal working range. On tuned circuits, temperature changes affect the frequency at which they operate, and manufacturers often publish figures of what one can expect.

For example, in the ICOM IC–R71 (communications receiver) stability is quoted as less than 500 Hz in the range of 0 deg. C to +50 deg. C. Also, less than 200 Hz between 1 minute and 60 minutes after switching on, and less than 30 Hz after 1 hour.

Selectivity and sensitivity

Selectivity is the ability to discriminate between stations operating on frequencies close together, and the sensitivity of a radio set refers to its ability to pick up weak stations. Better quality radios can be expected to respond to signals of 1μ volt or less, but at the other end of the scale a good set should also be able to cope with strong signals without overloading or distortion.

Some lower quality and early types of digital radios suffer from poor sensitivity due to the internal noise generated by their frequency synthesizers. In one or two sets that I have seen boasting full frequency coverage, SSB operation and many extras, the noise has been so bad as to make the set only useful for receiving strong local stations. Fortunately, most sets are not nearly as bad as this but some still leave room for improvement.

Extras

These days one hardly ever sees a truly basic radio set, especially at the better quality end of the market. It seems that manufacturers feel a need to strap on extra gadgets, but occasionally these extras become really useful and after a while it is hard to see how you managed without them.

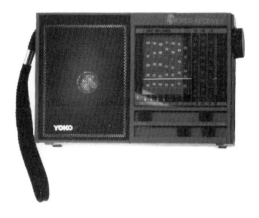

Photo 5.1 Battery portables. The Sony ICF 2001D (top right) is one of the better quality general coverage portables and, with a decoder, can be used to receive RTTY, weather fax, etc. It is ten times the price of the 'Yoko' analogue set (top left), which is suitable for broadcast station reception only. Like the Sony, the Sangean ATS–803A (bottom right) has digital frequency control and can be used to receive SSB signals, but is about one-third the price. Any of these sets can be protected from seawater by a plastic bag such as the 'Aquasac' (bottom left).

The frequency memories on digital radios are one example of this. Storing the operating frequencies of the stations you use regularly enables you to switch between them very rapidly. This is particularly useful if you are listening to a programme on one frequency and want to switch briefly to another, perhaps to check for some item that you are expecting to be broadcast.

Many sets are fitted with clocks and timers that can be set to turn them on (but not always off) automatically. If you are a regular listener to forecasts this can be useful, and even more so if you can arrange to record the forecast automatically too. Cassette tape recorders are sometimes built into radios but for one reason or another it is not always possible to set them up to record signals directly from the radio.

Internal ferrite rod antennas always have directional properties and when tuned to RDF beacons can be rotated to find the direction of the beacon. Some manufacturers have taken this a stage further and provided a purpose made RDF antenna as an accessory.

COMMERCIAL MARINE RADIO TRANSCEIVERS

Transceivers that are manufactured for working the marine bands differ from other transceivers in several important respects:

1. *Type approval*
All transceivers intended for use on the marine bands must be 'type approved'. This is the only kind of equipment for which licensing authorities will issue a ship licence, and so the use of any other kind of equipment on marine bands is illegal.

2. *Rugged construction*
Marine radios are expected to operate under harsh conditions and may have an important distress role. Because of this they are built to high standards of mechanical and electrical integrity.

3. *Simplicity of operation and frequency selection*
The controls on marine band transceivers are usually few in number and require no special skills to operate, the reason being that in an emergency any member of the crew should be able to use the set without too much difficulty. The methods used for changing frequency are particularly simple and generally use only simple switches or push button controls.

A characteristic of the marine bands is that they are divided up into numbered channels (see Reference Information, section II). Channel spacings are arranged so that radio operators have no need to tune to intermediate frequencies so that, in practice, changing to another frequency simply involves switching to another channel.

Marine VHF

Judging from outward appearances, the past 20 years may seem to have produced few fundamental changes in VHF transceiver design. Of course, there have been new fashions in front panels, some knobs have become buttons and sets are now smaller, but the basic controls remain the same. Installation work is also no different, needing the same antenna and power supply.

Inside the sets far more important changes have taken place as integrated circuits and frequency synthesizers have replaced discrete transistors and crystals. Whereas 20 years ago sets were sold with not much more than the two mandatory channels, 16 and 6, nowadays most sets are sold with 50 or more, although most people will make use of only a few.

Modern sets are also able to communicate better with weaker stations than their predecessors, which may be surprising since antennas remain the same and the maximum transmitter power is still 25 watts. The reason for this is that considerable improvements have been made in receiver sensitivity, but the point is only appreciated when you run two sets side by side and notice how many calls the older model fails to respond to.

Hand-held sets

These have been steadily decreasing in price and at the same time have acquired many of the features previously found on fixed sets, e.g. large numbers of channels, memory scanning, etc. Hand-helds also have much better mechanical protection and water resistance, though most stop short of being able to withstand even a quick dunking in sea water. This has always struck me as rather a a pity because often very little extra sealing would be needed to make them fully waterproof. Be that as it may, plastic bags made specifically for waterproofing hand-held sets have the additional benefit of making the set float.

The great advantage of hand-helds is that they can be used away from the boat or even in a life raft, should the need arise, which could be a reason for choosing a hand-held if your budget only extends to one VHF set. They are more limited in power and would not be expected to perform as well as a fixed set running its full 25 watts into a mast head antenna. Nonetheless, antenna height often counts for more and using a hand-held in conjunction with a masthead antenna would certainly improve its performance.

New facilities – digital selective calling

Digital selective calling is a service that is beginning to be introduced in several countries. The system being adopted was produced by the Italian company Cimat and, after an initial trial period at the Porto Cervo Coastal Radio Station, has been in widespread use in other Italian stations for a number of years. It has also been installed at Gibraltar's coastal radio station and will become operational throughout the UK, Greece, and Portugal.

The service is known by various names – Autolink in the UK and Seatel in Gibraltar, but it is intended to give marine VHF users (also MF users in the UK) the ability to call shore-based telephone subscribers direct. This can be done without intervention from the coastal station radio operator as all connections and billing will be carried out automatically by the station's computer.

For subscribers, the equipment needed is a small unit which is about the size of a fixed VHF set and includes a keypad and display window. Installation is straightforward and the unit can be coupled up to most types of transceiver. All that is required is a 12 volt power supply and connections to the microphone and extension speaker socket. An RS-232 serial port is also provided for those wishing to use FAX machines or exchange computer data.

Normal calls made through coastal stations can be picked up by anyone caring to eavesdrop on the band, but with this system, a speech scrambler can be switched in to preserve privacy. Also, by issuing subscribers with personal identification numbers (PINs) there is less opportunity for false billing or illicit use of the system.

The equipment manufactured by Cimat also has the capability of automatically connecting calls made to boats (digital selective calling) but unfortunately it would appear that many telecomunications authorities have no immediate plans to make use of the facility.

MARINE MF/HF TRANSCEIVERS

VHF communications are essentially line of sight only and any boat travelling more than 30 miles from the coasts and still wishing to enjoy the variety of services provided by coastal radio stations must turn to single side band MF and HF. MF is useful for ranges up to 300 miles but, depending upon propagation, HF signals can reach anywhere in the world.

No longer are these transceivers the size of a 'fruit-machine' nor do they demand vast amounts of power. The ICOM IC–M700 (shown with its antenna tuner) in Photo 5.2 covers both HF and MF and is fairly typical of current models. Its brief specification is as follows:

Transceiver dimensions: 124 mm H × 297 mm W × 379 mm D
Receiver coverage: 1.6 MHz to 23.9999 MHz
Transmit frequencies:
    2.0 to 2.9999 MHz
    4.0 to 4.9999 MHz
    6.0 to 6.9999 MHz
    8.0 to 8.9999 MHz
    12.0 to 13.9999 MHz
    16.0 to 17.9999 MHz
    22.0 to 22.9999 MHz

Radio equipment

Transmission modes: J3E, R3E and H3E (see p. 157)
Maximum power output: 150 watts PEP (SSB)
Power supply: DC 13.6 volts (±15%)
Current: 25 amps maximum (usually less)

The above transmit frequencies are those covered by the set in its standard form. However, with a small internal modification this set can be made to transmit amateur frequencies.

The receiver section of the IC–M700 gives good quality general coverage over a wide range of frequencies and so can be used for listening to short-wave broadcast from amateur stations. Also, with a suitable decoder it can be used to receive weather fax pictures and, because the transceiver is capable of semi-duplex (split frequency) working, it can be used for sending and receiving commercial telex.

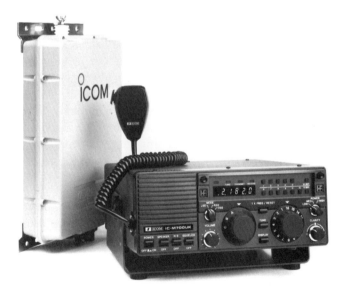

Photo 5.2 The MF/HF Marine band ICOM IC–M700 transceiver, together with its matching automatic antenna tuner. (*Courtesy of ICOM (UK) Ltd*)

AMATEUR EQUIPMENT

For many years the amateur equipment market has been dominated by three Japanese names – Icom, Kenwood and Yaesu. Western-based companies are beginning to make significant inroads but have tended to concentrate on specialist parts of the market. On the other hand, the Japanese have produced a whole range of equipment and accessories for frequencies from MF to UHF.

The main differences between transceivers built for amateurs and those intended for professional marine users stem from there being no need for manufacturers to submit amateur equipment for type approval. This gives rise to large variations in the quality of amateur gear but, at the better quality end of the market, differences between an item of amateur equipment and its marine equivalent are small. The most significant differences are as follows:

1. *Method of frequency adjustment*

Though some amateur bands are subdivided into channels (see Reference Information, section II) those below 30 MHz are not. Contacts can be made on any frequency within non-channelized bands and, to avoid interference from adjacent stations, operators sometimes make small frequency changes. To allow this, amateur transceivers are generally fitted with frequency controls which allow them to be continuously tuned across the band. Usually this is a large knob, rather like the tuning control on analogue receivers and is particularly useful in searching for stations for which the exact frequencies are not known.

2. *Extras and accessories*

Quite unlike marine band transceivers, some amateur equipment absolutely bristles with knobs, dials and all manner of gadgets. Manufacturers make no concessions in simplifying its operation and extras may include:

> Multiple memories for frequency storage
> Audio and intermediate frequency filters
> Speech compressor
> Integral antenna tuner
> Coverage of VHF and UHF bands on MF/HF sets
> Computer control interface
> Frequency scanning
> Spectrum analyser

Radio equipment

Photo 5.3 The Yaesu FT–747 amateur transceiver. (*Courtesy of South Midlands Communications Ltd*)

Once you have read and digested the instruction manual, many of these extras are undoubtedly most useful, especially those at or near the top of the list. None, though, are essential, and for maritime use my preference would be to choose equipment with a bare minimum of extras. Fortunately there are a few available, including hand-held VHF sets as well as fixed VHF and MF/HF sets, which are intended for mobile use.

The Yaesu FT–747 is one example of a particularly simple MF/HF amateur transceiver. The brief specification is as follows:

Transceiver dimensions: 93 mm H × 238 mm W × 238 mm D
Receiver coverage: 0.1 MHz to 29.9999 MHz
Transmit frequencies: 1.5 to 1.9999 MHz
(UK spec.) 3.5 to 3.9999 MHz
7.0 to 7.4999 MHz
10.0 to 10.4999 MHz
14.0 to 14.4999 MHz
18.0 to 18.4999 MHz
21.0 to 21.4999 MHz
24.5 to 24.9999 MHz
28.0 to 29.9999 MHz
Transmission modes: A1A, J3E, A3E (see p. 157)
Maximum power output: 100 watts PEP (SSB or CW)
Power supply: DC 13.5 volts (± 10%)
Current: 19 amps at 100 watt output

As with the ICOM IC–M700, this set is also capable of semi-duplex working and can be used for weather fax, AMTOR, packet and other data modes.

Emergency use of amateur equipment on marine bands

Many VHF and MF/HF amateur transceivers are quite capable of transmitting on marine band frequencies and often a small internal modification is all that is needed to achieve this. However, under normal circumstances, use of non-approved equipment on marine band frequencies is *illegal*. In an emergency which threatens life or the safety of the vessel this could provide a useful additional means of calling for assistance.

The scope for home-made equipment

There is nothing that can match the sense of achievement that comes with having built your own gear, but this apart, there are other, more practical, benefits. If your 'all bells and whistles' factory-built set breaks down, even if it is still under guarantee, finding someone to fix it when the nearest agent is 1000 miles away can be frustrating, to say the least. No radio is guaranteed not to break down, but if you built it yourself, the person best qualified to fix it is always close at hand.

At first sight it may seem that the odds are stacked against those who would build their own radios. On the marine bands, such attempts are discouraged by type approval requirements. For radio amateurs, home construction has always been part of the scene but it is unfortunate that the move towards SSB has produced a current generation of equipment of such complexity as to discourage all but the most ardent and resourceful constructors.

In spite of this rather negative scenario, there can still be considerable scope for home constructing, particularly if you lower your sights a little. Setting all band, all mode transceivers and receivers aside, and confining your attention to just one or two frequency bands makes home construction a much more realistic goal. This simplification need not mean making great sacrifices on performance, but can bring vast reductions in overall complexity and costs. As a result, there are many designs for simple pieces of radio equipment with a useful part to play afloat which can be assembled by anyone with average practical skills and resources.

A good way to start, particularly for those who may be apprehensive about home construction, is to assemble a piece of test equipment or even a simple receiver. It is worth having a go at battery chargers, audio filters (see below), dummy loads, SWR/

power meters, noise bridges and low-power antenna tuners (see Chapter 7). In all these cases relatively few components are involved and the capital outlay is quite small.

Quick success with any home-built project is more likely if you can start off with a well-tried design and, certainly, there is no shortage of these to choose from. Radio and electronic magazines regularly publish articles describing all manner of things for home constructors but, whatever the source of your design, it is important to check it thoroughly for any parts that might be critical or difficult to obtain, and for any 'experimental' aspects which could give difficulties.

One way to be reasonably sure of getting a proven design is by starting from a kit of parts. All manner of equipment can be bought in kit form, from simple test meters to quite complex transceivers, and this can certainly save much of the frustration of having to shop around for parts.

When looking at kits, find out exactly what parts are supplied. Some kits include all electronic components along with the case, knobs and every last nut, bolt and washer. Conversely, there are others that keep costs down by leaving out some of the hardware and, sometimes, the more readily obtainable electronic parts. Another point to check is the tools needed for assembly. In most cases you will only need hand tools, such as a soldering iron, screwdriver and pliers, but one or two more complex designs require alignment with electronic test instruments.

The following two examples of a single band receiver and low power transmitter show the kind of items that might be built at home. As far as kits are concerned, these rate about medium complexity and need no special tools for assembly.

SINGLE BAND DIRECT CONVERSION RECEIVERS

A general coverage communication receiver is an ambitious project for anyone to undertake but building a receiver for a small part of the HF spectrum is far less daunting. If, for example, you wanted to listen in to the amateur marine nets, a receiver that covered 14.000 MHz to 14.35 MHz (the 20-metre band) would get most of them.

Almost all nets use SSB and the simplest kind of radio able to resolve this would be a direct conversion receiver. These sets contain a tunable oscillator which is tuned to a frequency close to that of a desired station. The exact frequency of the oscillator is set so that when it is added to the incoming frequency, the resulting signal is of an audio frequency that can be filtered, amplified and sent to a loudspeaker or headphones.

There are theoretical limitations on the performance of these sets and one is that they are sensitive to interference from stations located on both sides of the oscillator frequency. For example, if there is a station on 14.251 MHz and to receive it the oscillator is tuned to 14.250 MHz, any strong interference on 14.249 will also be resolved. None the less, in practice, direct conversion receivers can give surprisingly good results and often out-perform many of the cheaper general-coverage portable sets, inspite of their more sophisticated superhet design.

Several successful designs for direct conversion receivers have been published by the RSGB and others. Typically, they use no more than a few dozen components and might be got working within an evening or two. At the time of writing, the cost of a kit was between £18 and £50 depending on how much it contained.

LOW-POWER MORSE TRANSMITTERS

Two advantages of radio Morse are that it can be produced using extremely simple equipment and can often be used successfully to communicate under conditions of heavy interference and marginal propagation. Also, when it comes to communicating over long distances, having a high transmitter power does not count for everything, as the following example should show.

Most communications receivers have a meter for measuring the strength of incoming signals. This is calibrated in 'S' points (calibrations vary but 1 S point ≈ 3 dB) and fair signals are reported as being strength S5 (see Reference Information, section I) whilst exceptionally strong signals are S9. Since a 50 per cent reduction in power corresponds to a drop of 1 S point, such a reduction on an exceptionally strong signal would still place it in the strong signal bracket. Even with a 75 per cent drop it would still be classified as a moderately strong signal or, put another way, it takes a three-fold power increase to turn an S4 – a fair signal – into an S6 – a good signal.

Simple, low-power Morse transmitters can be amazingly effective in kit form and their cost and complexity is similar to that of a direct conversion receiver. Photo 5.4 shows an example of a 12-watt transmitter (supplied by C. M. Howes Communications) which has proven its worth at sea on several occasions for

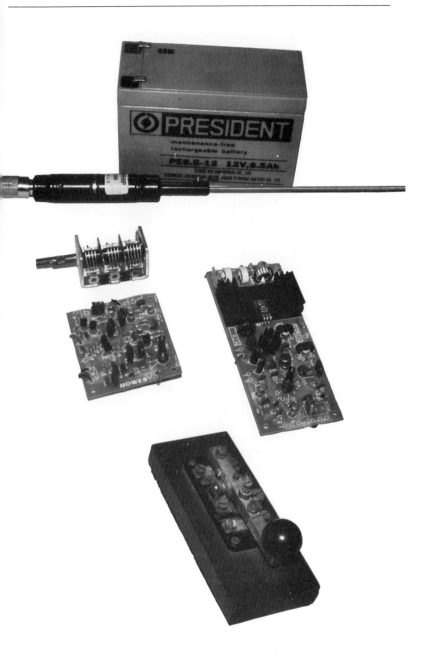

Photo 5.4 Parts for a kit-built low-power Morse transmitter. The two circuit boards are the transmitter (right) and variable frequency oscillator, along with its variable capacitor. Also included are a 12-volt rechargeable lead/acid cell, antenna (top) and wartime Morse key (bottom).

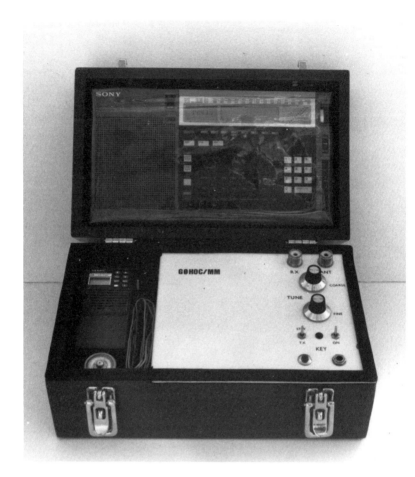

Photo 5.5 A standby low-power Morse transmitter and general coverage receiver.

communications of up to 2000 miles. This particular kit is supplied with all electronic components but constructors are left to provide their own case. This gives a great deal of flexibility so you can add shock or water protection as you feel necessary. You can also include whatever receiver arrangements you prefer, which could be a direct conversion receiver, as mentioned above, or you might consider using a good quality 'off the shelf' general coverage portable, as shown in Photo 5.5. Here the transmitter and radio are built into a waterproof wooden case which also holds a collapsible antenna, spare batteries and a miniature hand-held VHF transceiver.

6. Radio installation – Demands made on the boat

Anyone who has installed a piece of boat equipment will know that it is seldom as simple as the sales information might have you believe. Often the difficulties have a kind of knock-on effect and you end up working on parts far removed from the original job. Electrical installations are no exception but, apart from the mechanical problems of fixing units, transducers and running wires, there are also invisible effects upon the electrical system to be considered. The system must be able to provide enough of the right kind of power and, with radio equipment in particular, there is the special need to avoid interference. These are the subjects of this chapter, which begins with a review of installation requirements in general terms and concludes with some practical aspects of transceiver installation.

FINDING THE POWER

Some radios make no demands on the boat's power supply. EPIRBs and sets intended for emergency use only are often fitted with one set of batteries which remain in place for most of the life of the equipment, but these are the exception. In other kinds of radio equipment power will either come via a direct connection with the boat's electrical system, or from internal batteries if it is a portable item.

Batteries for portable equipment

Batteries for portable equipment come in a variety of shapes and sizes (see Reference Information, section IV), but, in addition to this, in recent years a wide range of types has become available. Their differing characteristics make particular types more suitable for some applications than others but, broadly, they can be divided into two groups, i.e. primary cells (non-rechargeable disposable types) and secondary cells (rechargeables).

PRIMARY CELLS (I.E. NON-RECHARGEABLE DISPOSABLE TYPES)

(a) *Zinc/carbon (Leclanché cells)*
Not too many years ago this type of cell, made by one manufacturer, dominated the market. Today they are still widely available and, though different qualities are made, when compared with other types they are the least expensive. They also store the least energy, but as with other kinds of battery, the exact amount that can be extracted depends very much upon how they are used.

Used continuously to drive a heavy current load (e.g. a transmitter, fast rewinding of a tape recorder, or an electric shaver) they will deliver less power than if used infrequently on a lighter load (e.g. a calculator, small portable radio with ear piece or LCD digital clock). A limitation here is their reduced shelf life and, if left to expire in a radio, corrosive products may leak out and cause damage.

(b) *Alkaline types*
These are used in much the same way as zinc/carbon batteries but may contain up to four times the energy content though, again, the exact amount will vary with the kind of load. Alkaline cells are best suited to applications where high currents are intermittently required.

(c) *Lithium/thionyl chloride*
The advantages of these cells are that they store a large amount of energy for their size and have very long shelf lives – manufacturers predict 10 years – and during this time they are expected to lose only 1 per cent of their charge per year.

Lithium cells have a voltage of 3.7, which is different from that of zinc/carbon or alkaline types. They are also manufactured in a different range of sizes and are much more expensive. Generally, lithium cells are only used in equipment where they can be installed on a more or less permanent basis, e.g. EPIRBs, distress beacons, and in computers or transceivers where they are used to provide memory backup power.

SECONDARY CELLS (RECHARGEABLE TYPES)

(a) *Maintainance-free lead/acid batteries*
These batteries are sealed in plastic containers and have a gel

electrolyte which eliminates the spillage problems of conventional lead acid cells. (An example is shown in Photo 5.4.) They can be charged or discharged in any position, and are well suited for use with portable equipment, but are not manufactured in sizes that enable them directly to replace zinc/carbon or alkaline cells.

Charging procedures are similar to those used on other types of lead acid batteries and similar equipment can be used, but of course the charge current will need to be adjusted to match the battery size.

Charging afloat is simplified if you can arrange to use batteries of a similar voltage to that of the boat's main supply but, again, voltage and current compatibility should be checked. When they are being charged correctly, unlike conventional lead/acid batteries, relatively small amounts of hydrogen gas are released into the atmosphere. However, considerable amounts may be released if they are over charged and vents are provided to allow for this.

When carrying out any battery charging, to avoid the risk of an explosion, it is important to ensure that the cells are freely ventilated and that care is taken to avoid sparks or naked lights in the immediate vicinity.

(b) *Nickel/cadmium (Ni/Cd) sintered cells*

These rechargeable cells come in all the familiar shapes and sizes of conventional zinc/carbon and alkaline cells but, in spite of a fairly high initial cost, they can be an economic alternative for radios, torches and a whole variety of portable equipment used afloat. Their great advantage is that, with a suitable charger, they can last for years and you have no need to carry stocks of batteries or worry about the problem of finding replacements.

At 1.25 volts per cell, their terminal voltage is slightly lower than that of disposable cells and this may cause problems with some voltage-sensitive equipment. These are usually electronic items and perhaps the only way of knowing if a particular piece is going to be susceptible is to check the manufacturer's information.

Advantages of Ni/Cd cells are:

1. Sturdy construction.
2. Can be stored for long periods without periodic recharging.
3. Can deliver high currents.
4. A life expectancy of 700 to 1000 cycles is typical.

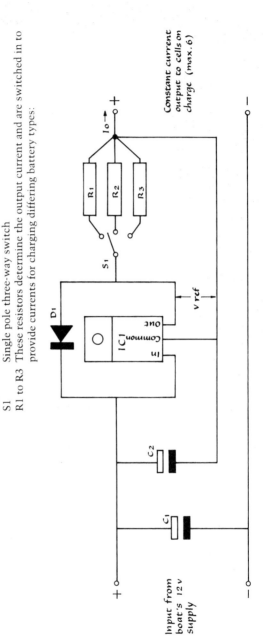

Fig. 6.1 Circuit diagram for an Ni/Cd battery charger, suitable for charging four cells in a series.

Key
IC1   Type 7805 (or similar) voltage regulator
C1   2200μF electrolytic capacitor. 25 volt
C2   1μF electrolytic capacitor. 25 volt (tantalum)
D1   1N4001 diode
S1   Single pole three-way switch
R1 to R3   These resistors determine the output current and are switched in to provide currents for charging differing battery types:

IC1 is a 5 volt (1 amp) regulator.
Output current is set by the value of R_1 to R_3 and values can be calculated from $R = \dfrac{V\,ref}{Io \times 1000}$ where Io = output current in milliamps
$V\,ref = 5$ (for a 5-volt regulator)

A characteristic of these batteries is their low internal resistance and this has several effects upon the way they are used. They should never be connected in parallel and, if they are short circuited, excessive currents can flow causing internal heating and, perhaps, serious damage. Apart from this, they are capable of delivering high currents and so are very suitable for loads such as portable transceivers, tape recorders, torches and electric shavers. Compared with other types of cells, they have a fairly high self-discharge (internal leakage) rate, which means that they are less suitable for electrical items that draw little current or stand idle for long periods. Examples here would be electric clocks, test meters and calculators (LCD types) emergency transceivers.

CHARGING NI/CD CELLS

Constant current charging is essential for Ni/Cd cells and the kind of battery chargers used for recharging lead acid batteries are unsuitable. Unfortunately, most chargers sold for Ni/Cd cells are designed to work off 100- or 240-volt a.c. mains supplies. This is fine if you are able to take them ashore for recharging but, afloat, the most efficient method of charging is from the boat's DC power supply. Fig. 6.1 gives a circuit diagram for a versatile charger that can be used to charge a variety of Ni/Cd cells from 12-volt supplies.

Battery type	R value	Output current	Approx. charge time
AA	82 ohms at 0.5 watts	60 mA	8 hours
C	22 ohms at 2.5 watts	227 mA	8 hours
D	10 ohms at 4.0 watts	500 mA	8 hours

On assembling the charger, the regulator IC should be bolted to a few square inches of copper or aluminium sheet to act as a heat sink. The resistors also dissipate a certain amount of heat and should be well ventilated.

Power for permanently installed equipment

VOLTAGE COMPATIBILITY

Thankfully, most boats use 12-volt electrical systems. This is a standard inherited from the road vehicle industry and is responsible for the emergence of the huge variety of appliances manufactured for this voltage. Some larger boats, following commercial vehicle standards, have 24 volts – this has the advantage of needing smaller cables to conduct the same power. However, users have a more restricted choice of electrical accessories which they may have difficulty in obtaining.

The umpteen million road vehicles with 12-volt electrical systems all produce electricity from engine-driven alternators and use lead/acid batteries for storage. A characteristic of these batteries is that, though they are nominally rated at 12 volts, their terminal voltage may rise to 15 volts when they are under charge, or fall below 11 when they are flat. Fortunately, most equipment intended to operate in vehicles or on 12-volt marine systems is manufactured to cope with these fluctuations, but in the case of items not primarily intended for these applications a check on the voltage compatibility is especially worthwhile.

Voltage fluctuations in marine systems (sailing boats in particular) are often greater than those found in road vehicles. This is because small loads, such as cabin lights, may be run for extended periods, thus flattening the battery and causing lower voltages than would normally be encountered in a vehicle. This problem is not usually too serious, but some radio equipment is particularly sensitive to low voltages, though it may function perfectly well aboard road vehicles. An example is the Yaesu FT-757GX transceiver which enjoys widespread popularity amongst boat owners, in spite of its 13.5 volt operating requirement.

Should you be stuck with equipment of lower voltage than the boat's supply, it may be possible to use a voltage adaptor (regulator) to make the conversion. In making these changes, some energy is always lost as heat and, where large currents are required, this loss may be unacceptable. With smaller capacity regulators, such as those sometimes used for portable radios where the current passed is no more than a few hundred milliamps, these losses are hardly noticeable.

FINDING THE ENERGY

On most small craft, and on sailing boats in particular, finding the energy to run electrical equipment can be a problem. Navigation lights, navigation equipment and engine starting have highest priority. Radio is, of course, also important, so when designing an installation from scratch or planning to install a new piece of

Radio installation – Demands made on the boat

equipment it is important to ensure that the power system can provide the energy needed, together with a protected reserve for high priority loads.

For radio and associated equipment, the table below gives a list of typical current demands. Those items included are typical of types in use today but, in general, older equipment may be expected to consume more current, and newer equipment less.

Equipment	Standby receive i.e. no signal (amps)	Receive (amps)	Transmit (amps)
100-watt amateur transceiver (Yaesu FT-757GX)	(Actual consumption depends upon how the set is used – see text)	2	19 max
25-watt VHF set (Aqua-marine 5600)	0.4 (These currents are reduced by 0.1 when front panel lights are turned off)	0.5	4.5
Hand-held VHF set with power saving (Yaesu FT-23R)*	0.019	0.15	1.5 (5-watt output)
Radio/cassette player (for broadcast reception)	0.2 to 2.8, depending upon output volume		
Portable computer (Amstrad PPC640D)	0.6 (2.2 with disk drive running)		
Decoders			
KAM (multimode)	<0.3		
PK-232 (multimode)	0.7		
MFJ (multimode)	0.4		
FAX-1 (WEFAX, NAVTEX & RTTY (decoder)	0.4		
Matrix Printer DCP-1	1.4 amps maximum		

* Some recent hand-held sets are fitted with a facility that allows them to use less power when standing by on a silent band. This is accomplished by switching the receiver off when no signals are heard. It remains off for a period of, say, 600 ms, after which it is turned on again. Then if no signals are received after some 300 ms, it is switched off and the cycle repeated. The output power of this kind of set is partially determined by the supply voltage. This particular model will accept voltages between 6 and 15 volts. At the lower end of the scale its maximum power output is reduced to 2 watts.

The current consumption of most items in the table on the previous page is quite modest. The HF transceiver would appear to be the exception but, in this case, peak currents are only drawn on SSB speech peaks, key-down CW or on some data modes. The electrical systems installed aboard most boats should have little difficulty in providing power for the occasional use of a transceiver. However, long periods on receive may cause problems, for instance if you are standing by waiting for a call, receiving WEFAX pictures or listening to broadcasts.

If you are compelled to operate on a tight energy budget, it will be necessary to make a note of the total number of ampere hours consumed and to ensure that the boat's charging system is capable of replacing them. Fitting larger batteries will not provide a long-term answer to energy shortages for, rather like having a bank account but no job, the benefits are short lived. Without adequate means of replacing energy consumed, sooner or later your batteries will go flat, a condition which will accelerate deterioration in lead/acid batteries.

Working within the limits imposed by the charging system is the only long-term solution. In the case of wind or solar power this may mean use is restricted to when the weather is sunny or blowing a gale. (Much of the power used by the word processor in writing this text was derived from the Levanter – the strong wind that frequently blows through the Straits of Gibraltar.) Of course, if you have shore power or are prepared to use the engine for charging, the problem will go away.

INTERFERENCE FROM ON-BOARD SOURCES

The effects of radio interference can be quite varied and occasionally equipment other than radios may be affected. Some effects are:

(a) Noisy radio reception over a wide band of frequencies.
(b) Radio reception on a single frequency is noisy or absent.
(c) Echo sounders show false readings or patterns.
(d) Weather fax pictures become cluttered by patterns or odd marks.
(e) Radar displays show sudden bright lines.*
(f) Autopilots behave erratically.*
(g) Satellite navigator display data is corrupted.*

  * On board transmitters are likely causes of these effects.

Radio installation – Demands made on the boat

Natural sources of interference include thunderstorms and atmospheric static and, apart from disconnecting the antenna, there is not much we can do about them. This also applies to most man-made sources in the outside world but often the most troublesome types originate from within the boat. In this case, once the exact cause has been located, it may be possible to effect a simple cure, but actually locating the source is often the bulk of the problem. As a starting point some common sources of interference are listed below:

(a) Alternators.
(b) Any form of electric spark (e.g. ignition systems on gas or diesel heaters, outboards or generators).
(c) Computers and other digital equipment (e.g. electronic typewriters).
(d) Dynamos or electric motors (especially with worn brushes or commutators).
(e) Echo sounders.
(f) Power inverters – including those fitted to low-voltage fluorescent strip lights.
(g) Mains voltage dimmers and electronic motor speed controllers (e.g. as fitted to sewing machines).
(h) Radar sets.
(i) Shore power connection.
(j) Switches, thermostats, and relay contacts.
(k) Bad electrical connections.
(l) Electrolytic corrosion between dissimilar metals.
(m) Rotating propeller shafts.

If you are lucky, the source of interference will be obvious and the problem may be solved simply by moving equipment around. If you are unlucky, tracking down the source of interference will require a careful and logical inspection of the boat's entire electrical system and, in the case of (k), (l) and (m) above, perhaps even mechanical and underwater parts as well.

Begin by progressively and systematically turning off parts of the electrical system whilst listening to the interference. If this fails to produce results, run the affected radio from a temporary set of leads connected directly to an isolated battery. If the interference is eliminated then a thorough check of the electrical system for bad connections is called for.

Most difficult to locate is interference which still persists when all electrical equipment is turned off and the main batteries isolated. The cause is most likely to be small electrolytic currents produced by wet dissimilar metals. Try looking for the source by moving the affected radio around to find the position in which interference is strongest, but if the source is below the waterline interference should disappear on hauling out.

Once you have located a source of interference the next question is what to do about it. To help understand the problem it is useful to divide interference into two types – connected interference (i.e. passed through power and connecting cables) and radiated interference.

Connected interference

This occurs as a small fluctuating voltage superimposed upon the vessel's DC power supply and becomes troublesome if picked up and amplified by radio equipment. (See Fig. 6.2.)

LOW FREQUENCY RIPPLE

Battery chargers with unsmoothed outputs are a common cause of this kind of interference which, with 50 Hz mains, may appear as a distinctive 100 Hz tone in radio loud speakers. Smoothing components usually include a large value electrolytic capacitor (several thousand microfarads) connected across the charger output and, possibly, a series choke, but these components are not often included. On the cheaper chargers they are left out for economic reasons but, on some more expensive electronic types, the addition of smoothing components may interfere with voltage regulation.

Wind generators can produce a similar effect, but in this case the tone frequency will vary with wind speed. On one boat I worked on the effect was picked up by the VHF transceiver but, although it could not be heard on receive, it seriously affected the transmitted signal and communications were lost as the wind gusted.

HIGH VOLTAGE SPIKES (TRANSIENTS)

High voltage spikes may be produced when switching inductive loads (e.g. pumps, motors, relays, solenoids, coils, etc.) and though they may be extremely short lived, such large voltages may be sufficient to damage sensitive electronic equipment. Failures of this kind can be especially troublesome to track down as the

Radio installation – Demands made on the boat

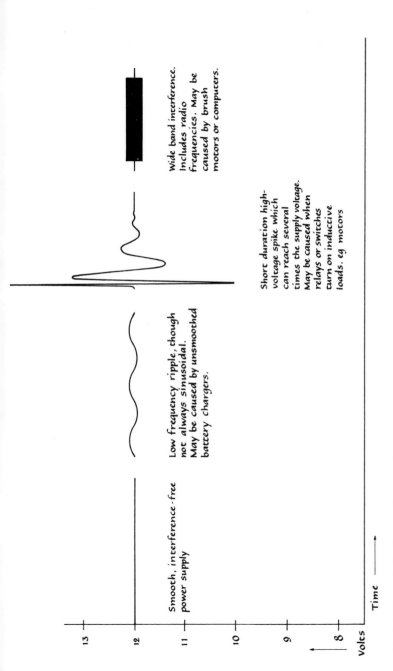

Fig. 6.2 Three kinds of power line interference.

effects are so brief, require specialist equipment to observe, and may be difficult to reproduce. Fortunately, manufacturers of marine electronics are well aware of just how electrically noisy boat electrical systems can be and the damage this can cause. As a result, most of the equipment on sale today includes circuits which are able to cope with almost all power supply noise, but this has not always been the case.

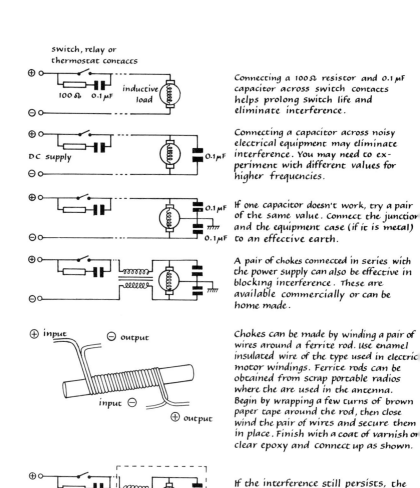

Fig. 6.3 Methods of reducing interference.

Even if high voltage transients are not a noticeable problem, some simple precautions to minimize transients are good practice and may avert a premature equipment failure. These include:

1. Replace brush gear on electric motors or dynamos before wear causes arcing and commutator damage.
2. Fit a capacitor and resistor across switch contacts which control inductive loads (see Figs 6.3 and 6.4). Pressure and float switches used to control water pumps, and relays controlling autopilot helm motors, all benefit from this treatment, which also extends switch contact life.

Radiated interference

Some types of interference are not constrained by connecting wires but pass freely into space, like other kinds of radio transmissions. The connecting leads of equipment producing radio frequency interference act as antennas, so allowing the interference to affect equipment some distance away. Often the noise is transferred between cables that run within the same group but it may also be picked up by other metal items (e.g. standing rigging, metal window frames, etc.) and reradiated into space. This characteristic adds further to the difficulties of locating the source of interference. This radio frequency noise can be produced in a variety of ways but aboard boats most likely sources are:

1. Any kinds of electric sparks, such as may be produced by electric motor/dynamo brushes, bells, buzzers or ignition systems.
2. Any kind of equipment in which currents are rapidly switched, e.g. alternators, power inverters and digital equipment (such as computers).

When installing new equipment the possibilities for interference can be reduced by taking care over the kind of cables that are used and the way in which they are laid:

1. Use screened cable for any interconnections to decoders, computers and accessories, and make sure that the screen is grounded at one end of the run. Grounding at both ends is not always a good idea as noise may be picked up by the loop that this creates.

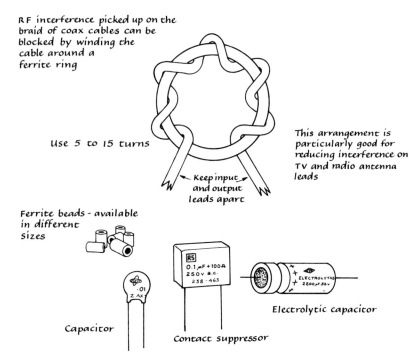

Fig. 6.4 A selection of devices for controlling interference.

2. Try to ensure that the positive and negative conductors (or, in the case of AC circuits, line and neutral) of each individual circuit are kept close together and not routed separately.
3. Try to keep any HF-transceiver wiring away from wiring for autopilots or electronic navigational equipment such as satellite navigators or Loran sets.
4. Make all interconnections as short as possible.
5. Try to keep antenna runs away from equipment control lines and/or interconnecting wires. If they need to cross, they should do so at an angle close to 90 degrees.
6. Ground leads should be as short as possible and earthing must be effective.

Dealing with this kind of interference, once it has become a problem, is often a case of trial and error. What works in one set of circumstances may not work in others but occasionally simply resiting affected equipment makes the difficulties go away. Fig. 6.4 gives more general ideas for tackling radio frequency interference.

NOISY COMPUTERS

Electrical noise in computers can be especially troublesome if the computer is to be used with a radio for reception of data transmissions or fax pictures. Points to check (listed in order) are as follows:

1. Use screened cables for all interconnections and make sure that one end of the screen is connected to an effective earth.
2. Thread ferrite beads along leads at the point of entry to the computer.
3. Try moving the computer. Using it the other side of a metal bulkhead may provide sufficient screening.
4. Try constructing an earthed metal screen around the computer – experiment with cooking foil.

Some users have successfully silenced their computers by metal-spraying the inside of the case. (See 'Reduction of RF breakthrough from the BBC Microcomputer', J. C. Worsnop, *Radio Communication*, Dec., 1987.) The results may be good, though this is no job for the faint hearted but a desperate measure requiring great care and confidence, and one that is certain to invalidate any warranty. First, the plastic case must be removed from the computer and cleared of all fittings. Those areas in which metal is not required should then be masked. Conductive aerosol paints are available for the job but best results will probably be achieved by entrusting the work to a specialist metal sprayer. The coating is connected to ground, but before assembly very great care must be taken to make absolutely certain that no electrical connections or components within the computer are able to make contact with the new metalwork.

Avoiding interference to other users of radio (and related equipment)

With any radio transmission there is no guarantee that it will only be picked up by equipment designed to receive it. Other radio users can suffer interference from transmitters which emit signals on the frequencies apart from those to which they are tuned. The cause may be bad design or maladjustment, but even perfectly adjusted transmitters can occasionally produce signals that 'break through' on other electrical equipment, particularly if it is close by. At sea, such problems are unlikely to occur but in close proximity to other boats, say in a marina, the situation is quite different. In cases like this the Ship Radio Licence only allows marine band calls to be made to port operations and similar services. However, in the absence of any local restrictions, the use of CB or amateur frequencies may be allowed but it is important to take steps to avoid interfering with other radio users, and with emergency services in particular.

Exceptionally strong radio signals can affect almost any kind of electrical equipment but perhaps most complaints arise from 'breakthrough' on domestic radios or television sets. In these cases there is no doubt that some (especially older) sets are unduly sensitive to strong, off-frequency signals; this is a difficulty that recently adopted manufacturing standards have done much to relieve. Nonetheless, even if the neighbour's TV set is unreasonably sensitive and picks up noise from every car or outboard that passes, if your transmissions blot out their favourite evening's viewing it could be hard to convince them that it is their set that is at fault.

Practical solutions to these problems often lie in a tactful and diplomatic attitude rather than in changes to equipment. However, by adopting good practices in installation and operation, such difficulties are less likely to occur.

For example:

(a) Ground the cases of all equipment which handle RF.
(b) Use short, direct ground leads.
(c) The use of braid breakers and toroids can help eliminate RF on the outside of coax screens (see Fig. 6.4).
(d) Once you have established a contact, reduce transmitter power to the minimum needed to maintain it.
(e) Disconnect any shore power connection and run on the boat's batteries whilst transmitting.

PROTECTION AGAINST LIGHTNING

Lightning strikes pass currents of around 2000 to 200,000 amperes and, though of only short duration, can still cause dangerous heating effects in materials through which they pass. If these are poorly connected metal parts or indifferent insulators, such as wet wood, such effects may be explosive and cause fires. Also, because of the large magnetic effects associated with currents of this size, fields within the boat will be substantially altered, and even if the

Radio installation – Demands made on the boat

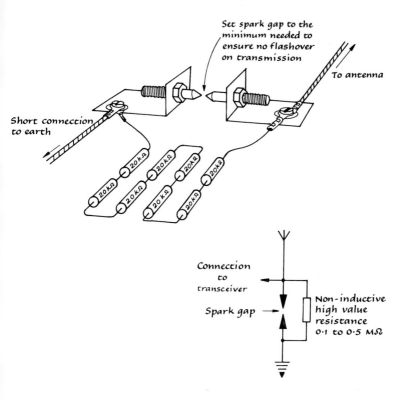

Fig. 6.5 High-voltage leakage and spark discharge path for antennas.

compass itself is unaffected its deviation may be radically changed. Other magnetic effects could include the erasure of cassette tapes and computer disk memory.

The probability of being struck by lightning varies considerably between different parts of the world but its effects are so devastating that provisions to limit them are a sensible precaution.

As far as radio equipment is concerned, it does not require a direct strike to cause considerable damage. Insulated metalwork, such as antennas, can acquire high voltages from static charges or strikes close by. Not only are these likely to damage any equipment to which they are connected but they are a danger to anyone touching them. In addition, on wooden or plastic hulls there is also an increased possibility of the boat's low voltage power supply lines acting as an antenna. In this way, large voltages may appear on otherwise isolated parts of the electrical system and so cause damage to any other electrical equipment aboard.

During a storm it is preferable to disconnect the antenna and power supply lines from all radio equipment and for the antennas to be connected to earth. However, there is a risk attached to handling antennas once a storm has actually started, but discharge of static electricity could be achieved by including the arrangement shown in Fig. 6.5.

HF TRANSCEIVER INSTALLATION

The performance of an HF transceiver is perhaps influenced more by its installation than by any other item of boat radio equipment. Its location and the way in which it is set up can make the difference between straining to hear some weak station and being able to hold a comfortable conversation.

Perhaps the most critical part of any transceiver installation is the antenna, so much so that the whole of the next chapter is devoted to the subject, but here we look at some important points in planning an installation.

SITING

In choosing a site for an HF transceiver, apart from the obvious need to find somewhere from which it can be operated conveniently, the site must also provide thorough protection from splashes of salt water and adequate ventilation. Even modern transceivers dissipate a fair amount of heat, especially when transmitting at full power. Older valve sets generate even more, and in either case, if the heat is not allowed to escape freely, the temperature rise is certain to damage internal components. Many sets are fitted with fans to assist internal cooling and, in choosing a site, it is vital to ensure that air can circulate freely around all parts of the unit.

POWER SUPPLY CONNECTIONS

Many transceivers are particularly sensitive to power supply voltage variations (see the section on voltage compatibility, p. 85). Transmitting from a low voltage supply can result in a distorted, unintelligible signal, sometimes referred to on SSB as 'FM-ing'. The problem is made worse if the power supply cables are too thin and/or too long as the high currents drawn on SSB speech peaks lead to large voltage losses.

Cable sizes of 6 or 10 sq mm are fairly typical but the actual size and length used should be chosen to keep the operating voltage within the transceiver manufacturer's range, even when transmitting at full power.

RF GROUNDING

In this context the term 'grounding' refers to the provision of a path for RF current to earth, which in the case of small boats usually means the surrounding sea. The term is not to be confused with connections to the negative supply terminal of the battery.

With most installations, the quality of the RF grounding has a significant influence upon overall performance and it is well worth taking pains to ensure that it is effective. On metal boats the ideal RF ground is the hull itself but in the case of wooden or plastic hulls, the solution may be less obvious. What is required is a large area of metal – at least 1 sq m, in close proximity to the water. Perhaps surprisingly, physical contact is not essential. In the case of fibreglass hulls one answer is to use a sheet of copper laid against the inside of the hull below the water line. Its effect is to form a capacitive connection through the hull, through which RF current is able to pass, though of course any DC would be blocked. On new fibreglass hulls quite a neat result can be achieved by a sheet of perforated copper foil bonded beneath the last layer of matt, though few boat builders have the foresight to do this.

Other possible means of achieving an effective RF ground are as follows:

(a) *Connection to an external metal keel or at least 1 sq m of metal (usually copper) attached to the outside of the hull.*
A metal rudder may also provide sufficient area, but do not be tempted to rely upon the propeller, skin fittings and stern gear.
(b) *Connection to an encapsulated metal keel.*

In this case connection to the sea water is capacitive and, again, it is important to have sufficient area of metal in close proximity to the water. For this reason ballast keels that consist of separate blocks of metal are unsuitable, and this also includes keels made from shot or steel punchings.

(c) *Proprietary brands of earthing plates (e.g. 'dynaplates') bolted to the outside of the hull.*

These proprietary earthing plates are usually constructed from a fused mass of tiny metallic spheres. With this kind of construction the area of metal in contact with the water is far in excess of that which might be estimated from the overall dimensions of the plate. Smaller models carry the equivalent of 1 sq metre of copper, and larger types intended for lightning protection are also available.

To work effectively it is important that they maintain a good contact with the water and to do this they should not be painted with antifouling. Regular cleaning on haul outs is essential and some users have reported that vinegar, or other mild acid, is effective in removing the scaling that sometimes occurs.

Connection between RF ground and transceiver

As mentioned at the beginning of this section, the object of the RF ground is to provide a low electrical impedance connection to the sea. This is made up partly by the impedance between the grounding plate (keel, or whatever) and the sea, and partly by the impedance of the connection between the plate and the transceiver. To achieve best results the route used for this connection should be as direct as possible and made of low RF impedance material. The traditional material is 25 mm × 1 mm copper strip, a better choice than conventional multistranded cable.

7. Antennas for use afloat

All radio equipment needs some kind of antenna to receive or send information to the outside world. In the case of small broadcast receivers the antenna may be hidden inside the case but often they are more conspicuous and appear in a range of shapes and sizes. It may be a simple, straight length of wire or a more complex structure, perhaps protected inside a plastic mushroom, tube or dome. In the case of transmitters, the design and location of the antenna will have a most profound effect upon the overall performance of the equipment and, with sets operating at frequencies below the VHF, antenna arrangements are usually in the hands of the person installing the set.

This chapter begins with a look at some basic factors affecting antenna design in general, then continues with some examples of simple but effective designs for HF transceivers. These same principles can then be extended to other frequencies and other types of equipment.

If you feel it unlikely that you will ever willingly undertake the task of constructing your own antennas, these basic ideas may still be useful. They can be of help in recognizing the frequency and hence the purpose of those you come across but, more importantly, in an emergency you will have some idea of how to make a replacement from materials that come to hand. Odd lengths of wire can often be cut to form a temporary antenna and though this may not be as effective as the original, it may provide the means of restoring communications.

SOME BASIC IDEAS

If you have little interest in mathematics, the relationships used to explain antenna design can appear daunting. The subject is

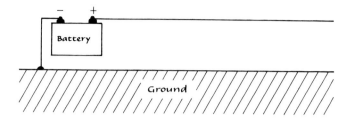

One end of a wire is connected to a DC source whilst the free end remains unconnected. As the net current flow is negligable, the wire appears to offer a high resistance, but all points along its length are at the same voltage.

Fig. 7.1(a) A DC supply connected to an open wire.

If the battery is replaced by a low frequency source of alternating current, provided that the wire is kept short, the voltage at any point along its length will follow that of the supply.

Fig. 7.1(b) A low-frequency AC supply connected to an open wire.

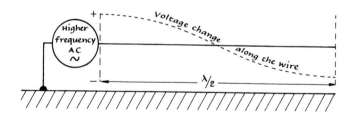

Current takes a finite time to travel the length of the wire. As the frequency or length of the wire is increased, the voltage at various points along its length may differ from that of the supply. The above diagram shows the special case when the length of the wire (metres) = ½ x 300/frequency (MHz). Here, the curve (superimposed on the wire) shows that the instant the supply voltage reaches a maximum, the voltage at the far end is at a minimum.

Fig. 7.1(c) A high-frequency AC supply connected to an open wire.

complex and a thorough treatment would occupy several volumes the size of this one, but fortunately it is not necessary to understand the whole subject to be able to construct an effective antenna.

The first difficulty that many people have in dealing with currents at radio frequencies is that the intuitive understanding they have gained in working with DC or domestic mains appears to break down. For example, there might be a sound (DC) electrical connection between two points in an RF circuit, but if the wire connecting them is coiled, current flow may be blocked. On the other hand, the same current may appear to flow into well-insulated wires that do not seem to be connected to anything at all, only to turn up again in other metalwork some distance away. However, this is behaviour we are already familiar with from the last chapter and the sections dealing with interference.

Fig. 7.1(a) shows a wire with one end connected to a DC supply whilst the other remains suspended freely. At the moment of closing the switch, current is able to flow into the wire. However, as the wire becomes charged, the flow soon stops and, like any other open circuit, it behaves as a high resistance with no net flow of current either into or out of the wire.

If the DC supply is replaced with an AC, as in Fig. 7.1(b), provided that the frequency is low and the wire quite short, the arrangment again behaves like a very high resistance. As the polarity of the supply goes positive, current flows in a direction that makes the wire also become positive. Later, as the supply goes negative, so the current flows in the opposite direction and the wire takes on a negative charge. Current in wires travels at speeds close to that of light, so fast that the charge even at the end of the wire appears to follow the polarity of the supply exactly. In this case, although current may be flowing back and forth as the supply changes direction, there is no net flow of current into or out of the wire so, as in Fig. 7.1(a), the overall resistance is again high.

If the length of the wire or the supply frequency is increased the situation changes and the effects become more interesting. As these changes are made, the point is reached when even the speed of light is not great enough to allow the voltage on all parts of the wire to follow exactly that of the supply. In these cases voltage and current distributions vary along the length of the wire. Fig. 7.1(c) shows the voltage distribution in the rather special case when the length of the wire equals half the wavelength λ.

Under this condition, known as resonance, a standing wave

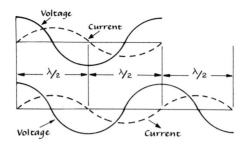

Fig. 7.1(d) Harmonic resonance on wires of other lengths.

pattern of voltage and current is set up along the length of the wire. The resistance, or impedance, that the antenna presents to the supply current (i.e. the radiation resistance) is now reduced and radio frequency energy is radiated into space. If small changes are made to the supply frequency or to the length of wire the resonance state is lost, and the antenna resistance increases. However, if larger changes are made, such that the length of wire becomes equal to a whole wavelength (or some other whole number multiple of a half wavelength) then resonance is re-established and the antenna resistance falls. Standing wave patterns for some of these cases are shown in Fig. 7.1(d).

Radiation patterns around a half-wave antenna

Most of the radiation from a half-wave wire antenna is produced in all directions at right angles to its length. Very little is transmitted along its axis. This is shown diagrammatically in Fig. 7.2, which shows a single large radiation 'lobe' surrounding the wire. At

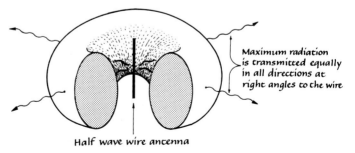

Fig. 7.2 Radiation lobes around a half-wave antenna.

higher harmonics, this pattern becomes more complex as extra lobes are formed, though the antenna is still generally less effective along its axis than it is at right angles. When this type of antenna is connected to a receiver, stations situated in the direction of the lobes may be expected to be received more strongly than those along the length.

PRACTICAL ANTENNAS

Boats vary so extensively in hull construction, deck and equipment arrangements, type of rig, etc., that it is impossible to arrive at a single type of antenna that will give optimum results in all situations. Above all, it must not interfere with normal handling of ropes, rigging or sails and should be as far as possible from the influences of close or overshadowing metallic structures. In fact, the electrical environment in which an antenna is placed can have a profound effect upon its performance, and choosing a suitable site invariably requires care, imagination and a few compromises.

As a result, it is seldom possible to achieve in practice the perfect doughnut-shaped radiation patterns described in the last section. These may be distorted in such a way that, in certain directions at right angles to the wire, radiation is increased yet at others it is reduced.

These and other uncertainties make the construction of small boat antennas a rather inexact science for most home constructors. However, with a little thought it is usually possible to produce a reasonable antenna on one's first attempt, though a little experimental work is certain to improve the results. In the section that follows we look at some simple antenna arrangements from which it should be possible to find at least one that can be adapted to give reasonable results in most situations.

Connection between the transceiver and antenna

The cable connecting a transmitter to its antenna is known as the feeder. It may consist of a spaced pair of wires held apart by a plastic separator (i.e. a balanced feeder) but these are rarely used on small boats and coaxial (coax) feeders are more common. These consist of a central copper conductor surrounded by a plastic insulating material. This is enclosed within a circular braided copper sheath, which is itself enclosed within a protective plastic outer cover.

Coax cable is manufactured in many different grades, but one property used to distinguish between different types is the 'characteristic impedance'. This depends upon the size and spacing of the conductors and the nature of the material that separates them. The type used for the majority of marine HF and VHF applications has a characteristic impedance of 50 ohm, as most transceivers are designed to match this. A 75 ohm coax is occasionally used for particular types of antennas.

Table 7.1 lists some properties of common types of 50 and 75 ohm coax, from which it can be seen that as the working frequency increases so do the losses (i.e. attenuation). Those cables with larger overall diameters are usually described as low-loss types and are correspondingly more expensive, but their main benefits are of more significance on VHF rather than HF frequencies.

Table 7.1   Characteristics of common types of coax cables

Cable type	Diameter (mm)	Characteristic impedance (Ω)	Attenuation (dB/10 m)		
			10 MHz	100 MHz	200 MHz
UR M43	5.0	50		1.3	
UR M67	10.3	50		0.68	1.0
UR M70	5.8	75		1.5	
RG58C/U	5.0	50	2.0		3.1
RG59B/U	6.15	75		1.3	1.9
RG213/U	10.3	50	0.18	0.62	1.0
WESTFLEX 103	10.3	50		0.32	0.54

Coaxial feeders are destroyed by dampness creeping through the cut ends or through any damaged parts of the outer sheath. When installing them make sure that the ends are well sealed to prevent water damage, and that they are given extra protection from mechanical damage.

The half-wave dipole

This is one of the most popular and effective types of antenna and is made by splitting the half-wave of Fig. 7.1(c) (see p. 102) into two quarter-wave sections. In its simplest form it is essentially a single-frequency antenna, but nonetheless, in practice, in addition to its resonant frequency, it can also be effective over a small band of frequencies on either side, as well as on others that are harmonically related. (Though in this latter case it should no longer be thought of as a *half*-wave dipole.) For example, a dipole cut for 14.175 MHz may still be effective on most other frequencies within the 20-m amateur band as well as some in the 10-m band. Fig. 7.3 shows the constructional details of a dipole antenna, but in practice the actual lengths of each quarter-wave leg (L) need to be cut slightly shorter than the theoretical length. The length for a half-wave dipole can be calculated as follows:

1. Take half the theoretical wavelength $= \frac{1}{2} \times \left(\frac{300}{\text{frequency (MHz)}} \right)$.
2. Reduce this amount by 2.9% (velocity factor).

Table 7.2 Practical lengths for half-wave dipoles

Frequency total (MHz) dipole	Length of a half-wave (metres)†
1.8	80.92
*2.182	66.75
4.2	34.68
3.6	40.46
6.4	22.76
7.05	20.67
8.5	17.13
12.7	11.47
14.17	10.28
16.8	8.67
21.2	6.87
22.4	6.50
28.2	5.165
*156.8	0.929

* International distress and calling frequencies.
† Note that the length of each leg will be half this amount.

The type of antenna illustrated in Fig. 7.3 can give excellent results in practice, in spite of it having a number of theoretical imperfections (e.g. the impedance of a half wave of 73 ohms, which is a poor match with 50-ohm coax and most transceivers).

Half-wave dipoles can be quick to make and the materials are usually cheap, though there may be the problem of finding somewhere clear to locate it. To avoid interfering with sails and rigging it may only be possible to hoist the antenna whilst motoring, at anchor or in harbour, but the difficulties become more acute on smaller boats and with lower frequencies.

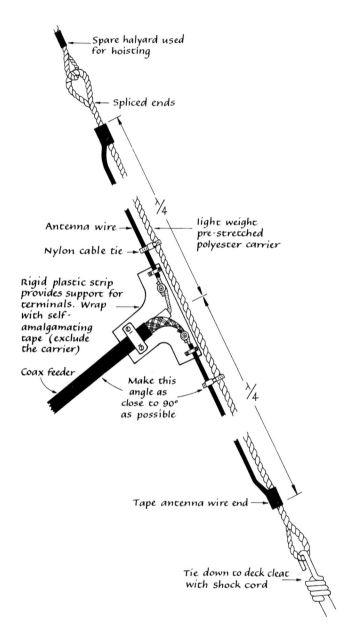

Fig. 7.3 Half-wave dipole construction details.

Antennas for use afloat

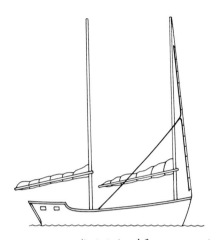

A temporary dipole hoisted from a spare halyard

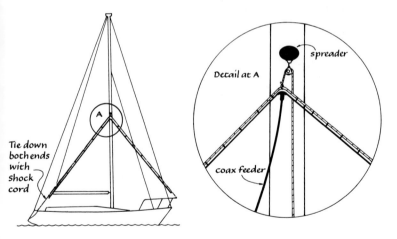

With dipoles that are too long to hoist from the mast head, satisfaction may be obtained by hoisting the centre and tying the ends down. Make the apex angle as large as possible (greater than 90°).

Fig. 7.4(a) and (b) Hoisting a dipole as a temporary antenna.

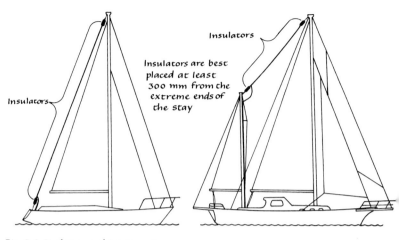

Fig. 7.5 Backstay and triatic antennas.

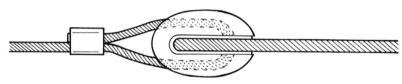

Pyrex or porcelain 'egg' insulators are traditional but will not withstand high rigging loads.

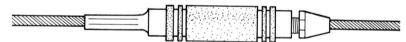

Plastic rigging insulators are available with a variety of end terminals. These include spade and eye fittings, swage and Norse or Stalock swageless terminals and can carry the full working load of the wire.

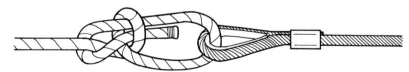

Polyester rope acts as a good insulator

Fig. 7.6 Antenna details.

Random wire antennas

On many boats, a satisfactory antenna can be made from an existing section of wire rigging. Such a wire can be fed from coax cable connected at one end, and to isolate it from other parts of the boat's rig, insulators will be required at both ends. In this case the effective length of the antenna is the length of the rigging between the insulators added to the length of cable connecting it to the set.

Any stay chosen to act as an antenna should not run close to other parts of the boat's metalwork and should be at least 8 m in length between the insulators. If it is shorter than this, tuning the lower frequency bands may be difficult, though performance in the middle and upper HF range could be satisfactory.

The main attraction of rigging antennas is that they are unlikely to interfere with the normal working of the boat but, since the main purpose of rigging is not to transmit/receive radio signals, it is quite likely that the length will not be related to any particular frequency that you might wish to use. If it is possible to arrange the insulators so that the antenna length is equal to half the wavelength (i.e. a 'practical' half wave length, including the length of the feeder) of a useful frequency, then so much the better.

In other cases, and if you wish to use the antenna on a range of bands, an antenna tuning unit will be required. This is a device able to alter the apparent electrical length of the antenna.

Antenna matchers/tuners

These are required to match the impedance of a random length of wire to the 50-ohm impedance of most transceivers. If this is not done, power will not be efficiently transferred between the transceiver and antenna, and the result could be damage to the output stage of the transmitter.

There are several proprietary types of matching units sold and the simplest (though not the cheapest) consists of a small box of concealed encapsulated components needing no external power supply and with no controls or adjustments. The box is simply connected to the antenna with as short a length of cable as possible. 50-ohm coax is used to connect the unit to the transceiver and connection to RF ground is also required but this is all.

These devices are claimed to prevent damage to the transceiver by always providing it with a 50 ohm impedance irrespective of the frequency and length of the antenna.

However, this does not mean that the device will make the antenna an effective radiator of any frequency delivered to it and in practice, best signals will be produced only at frequencies to which the antenna length is suited.

MANUAL TUNERS

In past years many antenna tuner designs have been published. Some are manufactured and sold as ready-made units, some as kits and others are suitable for complete home construction. Whatever type you choose, check to see that it can handle the maximum power you expect to use. It should be ruggedly built and constructed from good quality components.

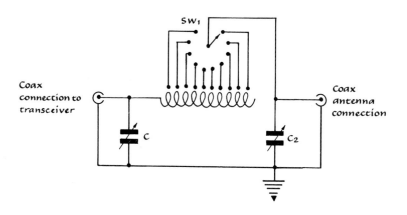

Fig. 7.7 A π section tuner

C1 350pF   air-spaced variable capacitor
C2 200pF   air-spaced variable capacitor. (This component should be designed for high-voltage working with a wide-plate spacing)
SW1   10-way single-pole switch
L1 15µH   In his book *Practical Wire Antennas*, John Heys suggests winding L1 from 20 turns of 14 swg wire (2 mm diameter) on a 76-mm diameter former with 10 turns wound at 4 turns/25 mm and 10 at 8 turns/25 mm. Alternate turns of the coil are tapped for connection to SW1.

The tuner illustrated in Fig. 7.7 is of an old and well-tried design. I have built it in various sizes to suit a 3 or 4 watt morse transmitter or large output transceivers. Power handling capacity is determined by the rating of the components used but it's always a good idea to use the best quality you can obtain. Connections should be as robust as possible and for strength and RF screening the unit should be assembled in a metal enclosure.

In using tuners of this type a certain amount of trial and error is needed to find the correct settings for the two capacitors and coil switch. The response of these controls is to some extent interrelated but, as a rough guide, good matching is usually found when they are adjusted to the point at which the strength of received signals is the strongest. This method, though, is not to be relied upon for transmitters and a check on the standing wave ratio (SWR) should always be made (see below).

SITING ANTENNA TUNERS

For convenience of operation, on many boats the antenna tuner is placed next to the transceiver. In these cases a fairly long run of cable is usually needed to connect it to the antenna but, as far as the tuner is concerned, this cable is also part of the antenna though, if coax cable is used, it is unable to function as such.

If it is not used, there is the possibility that radiated RF energy will cause feedback problems and difficulties with other equipment on board.

Ideally the tuner should be sited as close as possible to the point at which the feeder is connected to the antenna, and this generally means siting it outside in a protective enclosure, perhaps bolted down to the deck or attached to the pushpit or rail. If this is not feasible, consider mounting it in a stern locker or below deck but, whichever site is chosen, the antenna/tuner connection should be kept as short as possible.

SEMI- AND FULLY-AUTOMATIC TUNERS

Clearly, the disadvantage of a manual tuner is that, if the transceiver is some distance away, it will be inconvenient to adjust. These difficulties can be overcome by using a semi- or fully-automatic tuner.

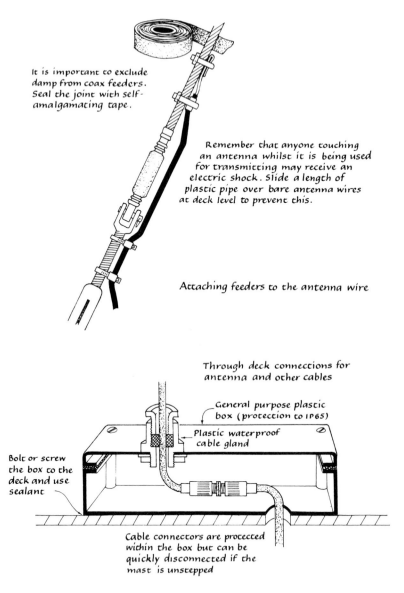

Fig. 7.8 Antenna details (cont.).

Semi-automatic tuners contain electrically operated switches which are used to change between the settings used on different bands. These are controlled by a remote switch which could be sited near the transceiver.

Fully-automatic tuners contain motorized controls which alter the tuner settings to match the band being used. (See the tuner in Photo 5.2.)

Whip antennas

On very small craft and power boats there may be no rigging suitable for supporting a long wire HF antenna. In these circumstances a whip antenna may provide the best solution but, when compared with the VHF equivalent, HF whips are often very much larger. At say 150 MHz (the mid-VHF) the length of a half-wave is 1 m, whereas at 15 MHz (the mid-HF) it is 10 times greater. Any whip built to these proportions would indeed be enormous and a correspondingly substantial support structure would be required to hold it in place. For these reasons, commercially made HF whips are usually considerably shorter and include, somewhere along their length, a 'loading coil', which makes up the difference between the physical length and required electrical length.

Whip antennas manufactured for amateur markets should be inspected very carefully if they are to be permanently installed aboard a boat. Some manufacturers make only feeble attempts at waterproofing or use materials that quickly corrode in a marine environment (e.g. chromium-plated steel or zinc/aluminium alloys). Beware, also, of those that contain heavy loading coils near the middle or top of the whip as these may impose unfair loading on the mounting structures as they swing about in a seaway.

VHF antennas

Because of the smaller size requirement, whip antennas are more common at VHF than HF. Those made specifically for marine band VHF are usually built to marine standards and prices are such that it is hardly worth considering making your own. It is worth noting, though, that should you suffer a dismasting, or some other situation that removes the main antenna, a makeshift dipole with each leg cut to 465 mm should produce a reasonable match into a marine VHF transceiver and an effective temporary replacement.

Because VHF range is essentially line of sight, the higher the antenna can be mounted the better. With sailing boats the obvious and most popular site is the masthead, where it is also clear of any influences from other metal structures. Unfortunately this is a prime spot on which navigation lights, wind instruments and

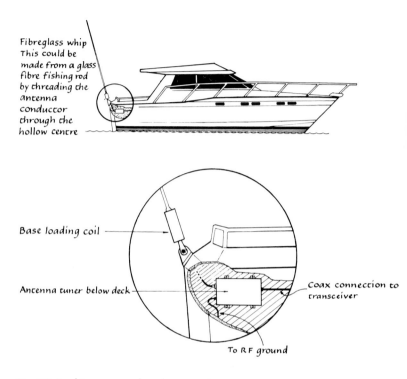

Fig. 7.9 A whip antenna aboard a power boat.

other antennas may also have a claim, though there are strict limits on sizes and weights that may be carried. Fig. 7.10 shows the acceptable separation between navigation lights and obscuring structures.

Antenna testing

Maximum power can only be transferred to an antenna system when its impedance is equal to (i.e. matched to) the impedance of the transmitter. If the transmitter/antenna match is poor, the antenna will only radiate part of the power supplied by the transmitter. The remainder is reflected back to the transmitter and may cause damage to the RF output stage (as mentioned earlier). For this reason, before transmitting on a new antenna system or after making adjustments it is most important to check that matching is correct.

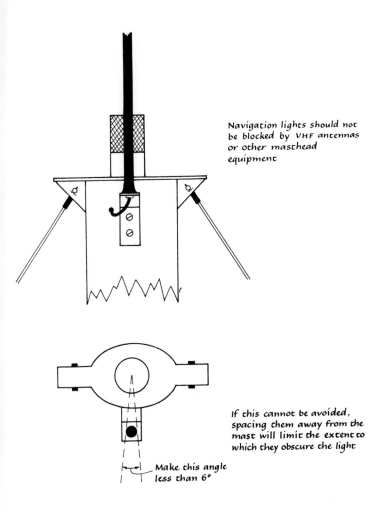

Fig. 7.10 Acceptable separation between navigation lights and obscuring structures.

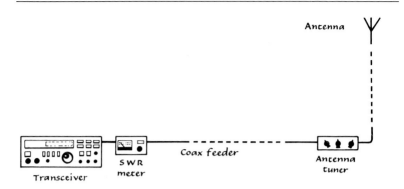

Fig. 7.11 Connections between a transceiver, SWR meter, tuner and antenna. The SWR meter checks the current/voltage ratio along the line.

STANDING WAVE RATIO (SWR) METERS

On a perfectly matched antenna system there should be no current/voltage standing waves along the feeder. This means that currents and voltages at all points along the line will be equal. The SWR meter is the instrument used to check this (Fig. 7.11).

Under ideal circumstances a SWR of 1:1 represents a perfect matching but, as this is seldom achieved, it is better to aim for a minimum. Ratios of 2:1 or even 3:1 are not too serious but action should be taken if values of 3.5:1 are reached. At this level the protective circuitry included in most modern transceivers begins to reduce output power.

SWR meters are often combined with a power meter and include a switch for changing between the two functions. Some transceivers have them conveniently built into the front panel. They are sometimes included on tuners as well but they are ideally connected close to the transmitter.

When tuning an antenna *use only low power* and make sure that you *listen on the frequency before transmitting* to ensure that it is not in use.

OTHER ANTENNA TEST EQUIPMENT

Dummy load
Sometimes it is necessary to test a transmitter under load but it would be wrong to do this while it is connected to an antenna, as the signal produced could well cause interference for other radio

users. The answer lies in the use of a dummy load, which is a high-wattage (often oil-cooled), low-inductive resistor that can be connected instead of the antenna. Its resistance (usually 50 ohms) is designed to match the transmitter output and metal screening ensures that no RF signal is radiated.

Noise bridge
A noise bridge is a useful and versatile piece of test equipment and can be used to measure several aspects of antenna performance without transmitting. In conjunction with a receiver it can be used to measure antenna impedance and resonant frequency or receiver input impedance.

Antennas for receivers

Design principles for receiver antennas follow similar lines to those for transmitters and one designed to transmit on a particular frequency will also be good for receiving the same, though the converse is not always true. In general, receivers are far less demanding of their antenna systems and mismatching will not cause damage, though the received signal strength may be reduced.

Short-wave and broadcast receivers are usually fitted with an internal ferrite rod antenna for reception on medium- and long-wave broadcasts. Most also have a telescopic antenna for VHF FM and short-wave reception but, with either type, reception of anything other than strong local stations is likely to be improved with an external antenna.

This is particularly true if you have a metal or ferro-cement boat, and in these cases, reception on portable radios below deck may be impossible due to the screening effect of the hull. Interference caused by noisy on-board electrical equipment can exacerbate the difficulties but, because it is further away from the source, an external antenna may again provide the solution.

ACTIVE ANTENNA

Active antennas are suitable for receivers only. Their name is somewhat misleading as it describes not an antenna but an amplifier. An active antenna is an extremely wide-band RF amplifier which is used to increase the signal from an ordinary antenna.

The strongest signals produced by an antenna will normally be at frequencies at which it resonates. However, a broad spectrum of other frequencies will also be present and an active antenna will amplify all of them, irrespective of frequency. Particularly if you wish to receive low frequency signals, and have no room for a long antenna, an active antenna may provide the answer.

Active antennas are quite small devices, generally fitted close to the antenna rather than the receiver. Some fit easily into a matchbox and, although a power supply is required, current drain is usually very low. There are several similar types on the market though, in kit form, circuits are usually simple enough for most people to assemble within a few hours.

Active antennas are not always the panacea for bad reception that their advertisers might have us believe. The main problems stem from the fact that, because they are wide-band amplifiers, they amplify all signals – the strong along with the weak. The effect may be that although the required signals are amplified so too are all background noises and interference. Also, if strong signals are present, even if not on the relevant frequency they may overload the RF section of the receiver, causing them to be heard on other parts of the band.

The solution to this difficulty lies in filtering the signal before it reaches the set. Inclusion of an antenna tuner of the type shown in Fig. 7.7 would do much to alleviate the problem. It would, of course, need retuning when the frequency is changed but, since it would not be used for transmitting, the tuner could be quite small and construction would be less demanding.

CONCLUSION

Just occasionally you may see aboard boats antennas which appear to defy all theoretical principles yet still give good results, so it appears that there is as much art as there is science to marine antennas. However, this does not mean that one should adopt a mystical approach and just throw up any odd length of wire and see if it works. Constructing and tuning an effective HF antenna can take time, but much frustration can be avoided if you begin with a well-tried design and are aware of some underlying principles. Even so, do not be content with the first result – be prepared to make adjustments and to look for improvements.

RADIO REFERENCE INFORMATION

I. Codes and procedures

MORSE CODE

Letter	Code	Letter	Code
A	.−	S	...
B	−...	T	−
C	−.−.	U	..−
D	−..	V	...−
E	.	W	.−−
F	..−.	X	−..−
G	−−.	Y	−.−−
H	....	Z	−−..
I	..	1	.−−−−
J	.−−−	2	..−−−
K	−.−	3	...−−
L	.−..	4	....−
M	−−	5	.....
N	−.	6	−....
O	−−−	7	−−...
P	.−−.	8	−−−..
Q	−−.−	9	−−−−.
R	.−.	0	−−−−−

Punctuation	Code
Full stop (.)	.−.−.−
Comma (,)	−−..−−
Colon (:)	−−−...
Question mark (?)	..−−..
Apostrophe (')	.−−−−.
Hyphen (-)	−....−
Fraction bar (/)	−..−.
Brackets − open (()	−.−−.
Brackets − close ())	−.−−.−
Double hyphen (=)	−...−
Quotation marks (")	.−..−.
Error	....

Accented letters	Code
à, á, â	.−−.−
ä	.−.−
ç	−.−..
è, é	..−..
ê	−..−.
ñ	−−.−−
ö	−−−.
ü	..−−

Readability of the code depends upon good timing and, though it may be sent at various speeds, it is important that the correct time relationships are maintained. The duration of the dash and the intervals between characters and between words should be related to the duration of the dot as follows:

Dash period	3 × dot period
Space between characters (within a single letter or number)	1 dot period
Space between letters or numbers	3 × dot period
Space between words	greater than 5 × dot period

INTERNATIONAL TELEGRAPH ALPHABET (ITA) No. 2

No. of signal	Letter case	Number case	\multicolumn{5}{c}{No. of element}				
			1	2	3	4	5
1	A	–	1	1	0	0	0
2	B	?	1	0	0	1	1
3	C	:	0	1	1	1	0
4	D	Who are you?	1	0	0	1	0
5	E	3	1	0	0	0	0
6	F	%	1	0	1	1	0
7	G	@	0	1	0	1	1
8	H	£	0	0	1	0	1
9	I	8	0	1	1	0	0
10	J	Bell	1	1	0	1	0
11	K	(	1	1	1	1	0
12	L	)	0	1	0	0	1
13	M	.	0	0	1	1	1
14	N	,	0	0	1	1	0
15	O	9	0	0	0	1	1
16	P	0	0	1	1	0	1
17	Q	1	1	1	1	0	1
18	R	4	0	1	0	1	0
19	S	'	1	0	1	0	0
20	T	5	0	0	0	0	1
21	U	7	1	1	1	0	0
22	V	=	0	1	1	1	1
23	W	2	1	1	0	0	1

No. of signal	Letter case	Number case	\multicolumn{5}{c}{No. of element}
			1	2	3	4	5
24	X	/	1	0	1	1	1
25	Y	6	1	0	1	0	1
26	Z	+	1	0	0	0	1
27	Carriage return		0	0	0	1	0
28	Line feed		0	1	0	0	0
29	Letters shift		1	1	1	1	1
30	Figures shift		1	1	0	1	1
31	Space		0	0	1	0	0
32	Not used		0	0	0	0	0

(1) Each code combination has two possible meanings. In any particular code group the correct meaning is identified by the shift code (i.e. 29 or 30) that precedes the group.
(2) In some countries, number case codes signals 6, 7 and 8 are used to represent characters of national significance, e.g. accented letters.
(3) American RTTY codes differ from ITA No. 2 in the following:

ITA No. 2	US RTTY
, (comma)	Bell
Bell	' (apostrophe)
"	+
;	=

THE STANDARD PHONETIC ALPHABET

Letter		Pronunciation	Letter		Pronunciation
A	alfa	ALfah	N	november	noVEMber
B	bravo	BRAHvoh	O	oscar	OSScar
C	charlie	CHARlee	P	papa	pahPAH
D	delta	DELLtah	Q	quebec	kehBECK
E	echo	ECKoh	R	romeo	ROWmeoh
F	foxtrot	FOKStrot	S	sierra	seeAIRrah
G	golf	golf	T	tango	TANGgo
H	hotel	hohTELL	U	uniform	YOUneeform
I	india	INdeeah	V	victor	VIKtah
J	juliett	JEWleeETT	W	whiskey	WISSkey
K	kilo	KEYloh	X	x-ray	ECKSRAY
L	lima	LEEmah	Y	yankee	YANGkey
M	mike	mike	Z	zulu	ZOOloo

THE STANDARD PHONETIC NUMERALS

Figure	Pronunciation	Figure	Pronunciation
1	wun	6	six
2	too	7	SEV-en
3	tree	8	ait
4	FOW-er	9	NIN-er
5	fife	0	zero

DISTRESS, URGENCY AND SAFETY SIGNALS

Distress

Use 2.182 MHz, or channel 16 VHF (156.8 MHz).
Speak *slowly* and *distinctly*

MAYDAY — MAYDAY — MAYDAY	Pronounced as in the French expression *m'aider*
THIS IS	Name of vessel 3 times
MAYDAY	Name of vessel once
MY POSITION IS	Lat./long. or true bearing and distance from a known position

The Distress signal is used to indicate grave and imminent danger to a ship, aircraft, vehicle or person requiring immediate assistance. Use of this signal imposes general radio silence, which is maintained until the emergency is over. If you hear a Distress call, write down the details. If no one else answers the call, you must do so. Then call a Coastguard station or a Coast Station with the details.

Urgency

Use 2.182 MHz, or channel 16 VHF (156.8 MHz).

PAN PAN — PAN PAN — PAN PAN	Pronounced as in the French expression *panne*
HELLO ALL STATIONS HELLO ALL STATIONS HELLO ALL STATIONS	Or specific station
THIS IS	Name of vessel 3 times

MY POSITION IS.............................. As for MAYDAY
.... Nature of emergency and
assistance required

The Urgency signal indicates a *very urgent* message concerning the safety of a ship, aircraft or other vehicle or the safety of a person, as in the case of serious injury or loss overboard. You would normally expect an immediate response. The message is in a similar format to a distress call. If you hear an Urgency call you should respond in the same way as for a Distress call.

Safety

The radio telephone safety signal consists of the French word *sécurité*, pronounced 'say-cur-e-tay', said three times.

It indicates a message about the safety of navigation generally, such as a drifting lightbouy, a wreck, a failed light or gale warnings. Such messages usually originate ashore, but they should be used by ships at sea to report a navigational hazard.

Safety messages are normally transmitted on working channels after an initial announcement on the calling channel.

Full details of all Distress, Urgency and Safety communications can be found in the *Handbook for Radio Operators* issued jointly by British Telecom International and Lloyds' of London Press.

PROWORDS (PROCEDURE WORDS)

All after	Used after the proword SAY AGAIN to request a repetition of a portion of a message.
All before	Used after the proword SAY AGAIN to request a repetition of a portion of a message.
Correct	Reply to a repetition of a message that has been preceded by the prowords READ BACK FOR CHECK when it has been correctly repeated.
Correction	Spoken during the transmission of a message means – an error has been made in this transmission. Cancel the last word or group. The correct word or group follows.
In figures	The following numeral or group of numerals are to be written as figures.

In letters	The following numeral or group of numerals are to be written in letters, as spoken.
I read back	If the receiving station is doubtful about the accuracy of the whole or any part of the message it may repeat it back to the sending station, preceding the repetition with the prowords I READ BACK.
I say again	I am repeating a transmission or portion, as indicated.
I spell	I shall spell out the next word or group of letters phonetically.
Out	This is the end of working to you. The end of work between two stations is indicated by each station adding the word OUT to the end of its reply.
Over	The invitation to reply.
Radio check	Please tell me the strength and clarity of my transmission.
Received	Used to acknowledge receipt of a message, e.g. YOUR NUMBER ... RECEIVED. In the case of language difficulties the word ROMEO is used.
Say again	Repeat your message or portion referred to, e.g. SAY AGAIN ALL AFTER ... or SAY AGAIN ADDRESS, etc.
Station calling	Used when a station receives a call which is intended for it but is uncertain of the identification of the calling station.
This is	This transmission is from the station whose call sign immediately follows. In the case of language difficulties the abbreviation DE spoken as DELTA ECHO is used.
Wait	If a called station is unable to accept traffic immediately it will reply to you with the words WAIT ... MINUTES. If the probable duration of the time exceeds 10 minutes the reason for the delay should be given.
Word after or word before	Used after the prowords SAY AGAIN to request a repeat of a portion of a radio telegram or message.
Wrong	Reply to a radio telegram that has been preceded by the words I READ BACK when it has been incorrectly repeated.

INTERNATIONAL Q CODE EXTRACTS

The value of these codes lies in that they are widely understood across language barriers. They can be used both as statements or as questions by adding a question mark (in radio telephony) and can be spelt out phonetically or sent in Morse. Some have even been incorporated as procedure codes in automatic telex machines.

QRA   What is the name of your vessel (or station)?
The name of my vessel (or station) is ...

QRB   How far approximately are you from my station?
The approximate distance between our stations is ... nautical miles (or kilometres).

QRG   Will you tell me my exact frequency (or that of ...)?
Your exact frequency (or that of ...) is ... MHz (or KHz).

QRH   Does my frequency vary?
Your frequency varies.

QRL   Are you busy?
I am busy (or I am busy with ... [name/call sign]).
Please do not interfere.

QRM   Is my transmission being interfered with?
Your transmission is being interfered with (strength rating 1–5).

QRN   Are you troubled by static?
I am troubled by static. (strength rating 1–5)

QRO   Shall I increase transmitter power?
Increase transmitter power.

QRP   Shall I decrease transmitter power?
Decrease transmitter power.

QRQ   Shall I send faster?
Send faster (... words per minute).

QRS   Shall I send more slowly?
Send more slowly (... words per minute).

QRT   Shall I stop sending?
Stop sending.

QRW   Shall I inform ... that you will call him on ... MHz (or KHz)?
Please inform ... that I am calling him/her on ... MHz (or KHz).

QRX   When will you call me again?
I will call you again at ... hours on ... MHz (or KHz).

QRZ   Who is calling me?
You are being called by ... on ... MHz (or KHz).

QSA	What is the strength of my signals (or those of ... [name/call sign])?
	The strength of your signals (or those of ...) is ... (strength rating 1–5).
QSB	Are my signals fading?
	Your signals are fading.
QSK	Can you hear me between your signals and, if so, can I break in on your transmission?
	I can hear you between my signals; break in on my transmission.
QSL	Can you acknowledge receipt?
	I can acknowledge receipt.
QSP	Will you relay to ... (name/call sign) free of charge?
	I will relay to ... (name/call sign) free of charge.
QSQ	Have you a doctor on board (or is ... [name of person] on board)?
	I have a doctor on board (or ... [name of person] is on board).
QSO	Can you communicate with ... (name/call sign) direct (or by relay)?
	I can communicate with ... (name/call sign) direct (or by relay through ...).
QSY	Shall I change to transmission on another frequency?
	Change to transmission on another frequency (or on ... MHz [or KHz]).
QTE	What is my true bearing from you? OR What is my true bearing from ... (name/call sign) OR What is the true bearing of (name/call sign) from ... (name/call sign)?
	Your true bearing from me is ... degrees at ... hours. OR Your true bearing from (name/call sign) was ... degrees at ... hours. OR The true bearing of ... (name/call sign) from ... (name/call sign) was ... degrees at ... hours.
QTH	What is your position in latitude/longitude (or according to any other indication)?
	My position is ... latitude ... longitude (or according to any other indication).
QTI*	What is your true course?
	My true course is ... degrees.
QTJ*	What is your speed? (Speed of a ship or aircraft through the water or air, respectively.)
	My speed is ... knots (OR kilometres per hour OR statute miles per hour).

QTL*	What is your true heading?
	My true heading is ... degrees.
QTM*	What is your magnetic heading?
	My magnetic heading is ... degrees.
QTR	What is the correct time?
	The correct time is ... hours.

* These codes have a similar meaning in the International Code of Signals.

THE RST CODE

This code is used for reporting the quality of signals in terms of their readability (R), strength (S) and tone (T).

Readability

R1 Unreadable.
R2 Barely readable, occasional words distinguishable.
R3 Readable with considerable difficulty.
R4 Readable with practically no difficulty.
R5 Perfectly readable.

Signal Strength

S1 Faint signals, barely perceptible.
S2 Very weak signals.
S3 Weak signals.
S4 Fair signals.
S5 Fairly good signals.
S6 Good signals.
S7 Moderately strong signals.
S8 Strong signals.
S9 Exceptionally strong signals.

Tone (i.e. Morse tone)

T1 Extremely rough hissing note.
T2 Very rough AC note. No trace of musicality.
T3 Rough, low pitched AC note. Slightly musical.
T4 Rather rough AC note. Moderately musical.
T5 Musically modulated note.
T6 Modulated note. Slight trace of a whistle.
T7 Near DC note. Smooth ripple.
T8 Good DC note. Just a trace of ripple.
T9 Purest DC note.

II. Frequency lists

BROADCAST BANDS

The following bands have been allocated by the ITU for use by broadcast stations:

Frequency (MHz)	Metre band (see **Notes** for area)	
2.300 to 2.495	120	(1)
3.200 to 3.400	90	(1)
3.900 to 4.000	75	(2)
4.750 to 5.060	60	(1)
5.950 to 6.200	49	(3)
7.100 to 7.300	41	(4)
9.500 to 9.775	31	(3)
11.700 to 11.975	25	(3)
15.100 to 15.450	19	(3)
17.700 to 17.900	16	(3)
21.450 to 21.750	13	(3)
25.600 to 26.100	11	(3)

Notes
1. Tropical bands – used only in designated areas.
2. Regional band – used only in Europe and Asia.
3. Bands used world-wide.
4. Not used in the Western hemisphere.

INTERNATIONAL BROADCAST STATIONS

Listening in to news broadcasts from other countries can give you a different perspective on world events, which can be important if the political future of the country you are about to visit is uncertain.

Frequency lists

Several publishers produce world-wide listings of broadcast station schedules but these have several shortcomings, particularly for people who travel widely.

Firstly, by selecting a suitable frequency and the use of beam antennas, broadcasts are targeted at particular parts of the world; outside these areas reception may be difficult or impossible. Second, some stations change their operating times and frequencies at least four times a year, for the following reasons:

1. To ensure that services to distant listeners are not disrupted by seasonal changes in propagation.
2. To avoid interference or deliberate jamming of the frequency.
3. To suit changing needs for the service.

The best way to obtain accurate broadcast schedules for a particular station is to request it by letter. Most are well aware of the difficulties listeners can have in obtaining this information and will freely distribute programme and frequency details.

Alternatively, if you manage to find one frequency on which a particular station operates, by listening in you may well hear details of others on which reception could be better.

The stations listed below have all been found to answer letters requesting schedules. Those frequencies listed alongside are used for English-language broadcasts, but for the above reasons some may be abandoned from time to time and resumed later:

Station	*Frequencies*	
Radio Australia,	7.205	15.320
GPO Box 428 G,	9.580	17.795
Melbourne,	9.655	21.740
VIC. 3001,	11.910	
AUSTRALIA.	15.240	
Österreichischer Rundfunk,	6.155	13.730
Shortwave Services, A–1136 Wien,	9.870	15.430
AUSTRIA.	11.780	21.490
	21.490	
Radio Canada International,	5.960	11.845
PO Box/CP 6000,	9.755	11.925
Montreal, H3C 3A8,	11.940	17.820
CANADA.	11.845	

Radio Praha,	5.930	13.715
Foreign Languages,	6.055	15.110
Vinohradska 12.	7.345	17.840
120 99 Praha,	9.630	21.505
CZECHOSLOVAKIA.	11.685	21.705
Danmark Radio,	11.720	15.165
KORTBØLGEAFDELINGEN,	11.740	17.720
1999 Kobenhavn V, DENMARK.	11.790	17.870
	11.930	
Deutsche Welle,	6.010	11.765
Listeners' Mail Department,	6.035	13.790
D–5000 Köln 1,	6.085	15.390
POB 10 04 44 WEST GERMANY	9.565	17.875
	9.700	21.600
Elliniki Radiophonia Tileorassi	7.430	15.625
(Hellenic Radio – Television),	9.395	15.630
Direction of Technical Services,	9.420	17.550
POB 60019,	11.645	
15310 Aghia Paraskevi Attikis,	12.105	
Athens, GREECE.		
Radio Budapest Broadcast,	6.110	11.910
Brody Sandor u. 5–7,	7.220	11.925
Budapest,	9.520	15.160
H–1800 HUNGARY.	9.585	15.220
Israel Broadcasting Authority,	7.460	11.605
Director of Overseas Services,	9.435	13.750
PO Box 1082,	9.930	15.615

Jerusalem   (Weather forecast for Mediterranean Coast
ISRAEL   and Gulf of Eilat broadcast at 0500 UTC)

Radio Nederland,	6.165	13.700
PO Box 222,	9.715	15.560
1200 JG Hilversum,	9.815	17.575
THE NETHERLANDS.	11.720	21.480
Radio Norway International,	5.980	17.740
Bj. Bjoernsons pl.1,	9.595	17.840
0340 Oslo 3,	15.165	21.705
NORWAY.	15.305	

Pakistan Broadcasting Corporation,	9.815	15.605
Broadcasting House,	11.610	17.895
Constitution Avenue,	11.570	21.480
Islamabad,	13.665	21.740
PAKISTAN.		
Far Eastern Broadcasting Company,	11.850	
Box 1,	15.480	
Valenzuela,		
Metro Manila		
PHILIPPINES.		
South African Broadcasting Station,	9.580	17.765
Piet Meyer Building,	9.615	21.590
Henley Road,	11.900	
Broadcasting Centre,	11.925	
Johannesburg 2000	15.230	
SOUTH AFRICA.		
Swiss Broadcasting Corporation,	9.725	13.685
CH–3000 Berne 15,	9.885	15.525
SWITZERLAND.	12.035	17.570
	13.635	
BBC World Service,	0.198	12.095
Transmission Planning Unit,	5.975	15.070
PO BOX 76,	3.955	15.310
Bush House,	6.045	17.710
London WC2B 4PH,	6.195	17.880
UNITED KINGDOM.	9.410	21.710
	9.750	25.750
United Nations Radio,	7.255	15.375
Room S–850,	9.545	15.565
New York 10017,	11.810	25.945
USA.	15.185	
Vatican Radio,	7.125	15.105
Ufficio Propaganda e Sviluppo,	9.650	15.190
00120 VATICAN CITY.	11.750	17.730
	11.955	17.870
	15.090	21.485

SINGLE SIDE BAND SIMPLEX FREQUENCIES FOR SHIP TO SHIP CONTACTS (USB).

Code	Carrier frequency (MHz)	Code	Carrier frequency (MHz)
	2.065	12A	12.353
	2.079	12B	12.356
	2.096	12C	12.359
4A	4.125	16A	16.528
4B	4.146	16B	16.531
4C	4.149	16C	16.534
6A	6.224	22A	22.159
6B	6.227	22B	22.162
6C	6.230	22C	22.165
		22D	22.168
8A	8.294	22E	22.171
8B	8.297		

HF CHANNELS FOR LONG RANGE TELEPHONE AND TELEX

United Kingdom
Portishead†

RADIO TELEPHONE DUPLEX CHANNELS (CARRIER FREQUENCIES)

CALL SIGN	ITU CHANNEL No.	SHIP TRANSMIT (MHz)	SHIP RECEIVE (MHz)
GKT20*	410	4.092	4.384
GKT22	402	4.068	4.360
GKT26	406	4.080	4.372
GKV26	426	4.140	4.432
GKU46*	816	8.240	8.764
GKT42	802	8.198	8.722
GKU49	819	8.249	8.773
GKU42	822	8.258	8.782
GKV46	826	8.270	8.794
GKW41	831	8.285	8.809
GKV54*	1224	12.299	13.146
GKT51	1201	12.230	13.077
GKT52	1202	12.233	13.080
GKT56	1206	12.245	13.092
GKV58	1228	12.311	13.158
GKV50	1230	12.317	13.164
GKW52	1232	12.323	13.170
GKT62*	1602	16.363	17.245
GKT66	1606	16.375	17.527
GKU61	1611	16.390	17.272
GKU65	1615	16.402	17.284
GKU68	1618	16.411	17.293
GKV63	1623	16.426	17.308
GKW62	1632	16.453	17.355
GKW67	1637	16.468	17.350
GKW60	1640	16.477	17.359
GKT18*	1801	18.780	19.755
GKU18	1803	18.786	19.761

* Denotes main frequencies. These are used for the broadcast of traffic lists and urgent messages. Vessels are requested to make their initial call on main channels.

CALL SIGN	ITU CHANNEL No.	SHIP TRANSMIT (MHz)	SHIP RECEIVE (MHz)
GKT76*	2206	22.015	22.711
GKU72	2212	22.033	22.729
GKU74	2214	22.039	22.735
GKU70	2220	22.057	22.753
GKV77	2227	22.078	22.774
GKV79	2229	22.084	22.780
GKX70	2240	22.117	22.813
GKU25*	2502	25.073	26.148

* Denotes main frequencies. These are used for the broadcast of traffic lists and urgent messages. Vessels are requested to make their initial call on main channels

RADIO TELEX CHANNELS (CARRIER FREQUENCIES)

CALL SIGN	ITU CHANNEL No.	SHIP TRANSMIT (MHz)	SHIP RECEIVE (MHz)
GKE2	2	4.173	4.211
GKE3	2	6.2635	6.315
GKE4	2	8.337	8.417
GKE5	2	12.4775	12.580
GKE6	2	16.684	16.8075
GKE7	2	22.285	22.377

These are main channels only. The code used is marine SITOR ARQ mode for telex traffic with ships. Traffic lists, weather warnings and general information is broadcast in FEC code.

United States

WOM – Fort Lauderdale, Florida

ITU CHANNEL No.	SHIP TRANSMIT (MHz)	SHIP RECEIVE (MHz)	ITU CHANNEL No.	SHIP TRANSMIT (MHz)	SHIP RECEIVE (MHz)
209	2.0315	2.490	1206*	12.245	13.092
221	2.118	2.514	1208	12.251	13.098
245	2.390	2.566	1209	12.254	13.101
247	2.406	2.442	1215	12.272	13.119
403*	4.071	4.363	1223	12.296	13.143
412	4.098	4.390	1230	12.317	13.164
417	4.113	4.405	1601*	16.360	17.242
423	4.131	4.423	1609	16.384	17.266
802*	8.198	8.722	1610	16.387	17.269
805	8.207	8.731	1611	16.390	17.272
810	8.222	8.746	1616	16.405	17.287
814	8.234	8.758	2215*	22.042	22.738
825	8.267	8.791	2216	22.045	22.741
831	8.285	8.809	2222	22.063	22.759

* Traffic lists are broadcast on these channels at odd numbered (UTC) hours, on the hour

KMI Inverness, California

ITU CHANNEL No.	SHIP TRANSMIT (MHz)	SHIP RECEIVE (MHz)	ITU CHANNEL No.	SHIP TRANSMIT (MHz)	SHIP RECEIVE (MHz)
242	2.003	2.450	1201	12.230	13.077
248	2.406	2.506	1202	12.233	13.080
401	4.065	4.357	1203*	12.236	13.083
416*	4.110	4.402	1229	12.314	13.161
417	4.113	4.405	1602	16.363	17.245
804	8.204	8.728	1603	16.366	17.248
809	8.219	8.743	1624	16.429	17.311
822	8.258	8.782	2214	22.039	22.735
			2223	22.066	22.762
			2228	22.081	22.777
			2236	22.105	22.801

*Traffic lists are broadcast on these channels at 00.00, 04.00, 08.00, 12.00, 16.00 and 22.00 (UTC).

WOO – Manahawkin, New Jersey

ITU CHANNEL No.	SHIP TRANSMIT (MHz)	SHIP RECEIVE (MHz)	ITU CHANNEL No.	SHIP TRANSMIT (MHz)	SHIP RECEIVE (MHz)
232	2.166	2.558	1203	12.236	13.083
242	2.366	2.450	1210	12.257	13.104
410	4.092	4.384	1211	12.260	13.107
411*	4.095	4.387	1228	12.311	13.158
416	4.110	4.402	1605	16.372	17.254
422	4.128	4.420	1620	16.417	17.299
808	8.216	8.740	1626	16.435	17.317
811	8.225	8.749	1631	16.450	17.332
815	8.237	8.761	2201	22.000	22.696
826	8270	8.794	2205	22.012	22.708
			2210	22.027	22.723
			2236	22.105	22.801

* Traffic lists are broadcast on these channels 00.00 (UTC) and all even numbered hours, on the hour.

MARINE VHF – CHANNEL NUMBERS AND FREQUENCIES

Channel No		Ship Transmit (MHz)	Ship Receive (MHz)	Function	Notes
00		156.00	156.00		1
	60	156.025	160.625	PO, PC	
01		156.050	160.650	PO, PC	
	61	156.075	160.675	PO, PC	
02		156.100	160.700	PO, PC	
	62	156.125	160.725	PO, PC	
03		156.150	160.750	PO, PC	
	63	156.175	160.775	PO, PC	
04		156.200	160.800	PO, PC	
	64	156.225	160.825	PO, PC	
05		156.250	160.850	PO, PC	
	65	156.275	160.875	PO, PC	
06		156.300	same	IS	
	66	156.325	160.925	PO, PC	
07		156.350	160.950	PO, PC	
	67	156.375	same	PO, IS	2,4
08		156.400	same	IS	
	68	156.425	same	PO	
09		156.450	same	PO, IS	
	69	156.475	same	PO, IS	
10		156.500	same	PO, IS	4
	70	156.525		IS	3
11		156.550	same	PO	
	71	156.575	same	PO	
12		156.600	same	PO	
	72	156.625		IS	4
13		156.650	same	PO, IS	
	73	156.675	same	PO, IS	
14		156.700	same	PO	
	74	156.725	same	PO	
15		156.750	same	PO, IS	
	75	Channel 16 guard band 156.7625 – 156.7875			
16		156.800	same		
	76	Channel 16 guard band 156.8125 – 156.8375			
17		156.850	same	PO, IS	
	77	156.875		IS	

Cont.

Continued

Channel No		Ship Transmit (MHz)	Ship Receive (MHz)	Function	Notes
18		156.900	161.500	PO	
	78	156.925	161.525	PO, PC	
19		156.950	161.550	PO	
	79	156.975	161.575	PO	
20		157.000	161.600	PO	
	80	157.025	161.625	PO	5
21		157.050	161.650 or same	PO	
	81	157.075	161.675	PO, PC	
22		157.100	161.700	PO	
	82	157.125	161.725	PO, PC	
23		157.150	161.750 or same	PC	
	83	157.175	161.775 or same	PC	
24		157.200	161.800	PC	
	84	157.225	161.825	PO, PC	
25		157.250	161.850	PC	
	85	157.275	161.875	PC	
26		157.300	161.900	PC	
	86	157.325	161.925	PC	
27		157.350	161.950	PC	
	87	157.375	161.975	PC	
28		157.400	162.000	PC	
	88	157.425	162.025	PC	

Functions: IS – intership; PO – port operations
PC – public correspondence

NOTES

1. UK channel 0 is used exclusively for communication between HM Coastguard and rescue services, e.g. lifeboats.
2. UK channel 67 is allocated for small craft safety and used by small craft and HM Coastguard.
3. Channel 70 is reserved for digital selective calling.
4. These channels may be used by ships, aircraft and land stations involved with search and rescue or pollution control operations.

5. UK channel 80 is allocated as a primary working channel for marinas. In case of overload channel M may be used.

In the UK, the private channel on 157.850 MHz, often referred to as channel M, is used for communication between yachts and marinas or clubs. The private channel on 161.425 MHz (M2) is allocated for use by yacht clubs.

Channel 6 is the primary intership channel. Channels 6, 8, 72 and 77 should be used in preference to others for this purpose.

RADIO TIME SIGNALS

Many countries transmit standard frequencies and time signals. Frequencies commonly used are 2.5 MHz, 5.0 MHz, 15 MHz and 20 MHz. The transmission format varies between stations and, in addition to the time signals, some include details of radio propagation, storm warnings and other coded information.

Station	*Call sign*	*Frequency (MHz)*	*Time (UTC)*
Argentina Buenos Aires	LOL2	5.0 10.0 15 Transmissions are on all frequencies simultaneously	11.00–12.00 14.00–15.00 17.00–18.00 20.00–21.00 23.00–24.00
Australia (Lyndhurst, Victoria)	VNG	4.5 7.5 12.0	09.45–21.30 22.45–22.30 21.45–09.30
Canada Ottawa	CHU	3.3 7.335 14.67	24H 24H 24H
France Paris	FFH	2.5	08.00–16.25 Not Sundays
Japan Tokyo	JJY	2.5 5.0 10.0 15.0	24H 24H 24H 24H

Radio Time Signals continued

UK	MSF	2.5	24H
Rugby		5.0	24H
		10.0	24H
USA	WWV	2.5	24H
Fort Collins,		5.0	24H
Colorado and		10.0	24H
Kekaha-Kauai,	WWVH	15.0	24H
Hawaii).		20.0	24H
		25.0 (WWV only)	24H
USSR	RWM	4.996	24H
(Moscow)		9.996	24H
		14.996	24H

CITIZEN BANDS – CHANNEL NUMBERS AND FREQUENCIES

27 MHz FM (F3E). Specification MPT 1320.
Maximum permitted power = 4 watts.

Channel	MHz	Channel	MHz	Channel	MHz
1	27.60125	15	27.74125	29	27.88125
2	27.61125	16	27.75125	30	27.89125
3	27.62125	17	27.76125	31	27.90125
4	27.63125	18	27.77125	32	27.91125
5	27.64125	19	27.78125	33	27.92125
6	27.65125	20	27.79125	34	27.93125
7	27.66125	21	27.80125	35	27.94125
8	27.67125	22	27.81125	36	27.95125
9	27.68125	23	27.82125	37	27.96125
10	27.69125	24	27.83125	38	27.97125
11	27.70125	25	27.84125	39	27.98125
12	27.71125	26	27.85125	40	27.99125
13	27.72125	27	27.86125		
14	27.73125	28	27.87125		

Channel 9 – Reserved for emergencies, though not officially monitored.

Channels 14 and 19 – These channels are used for calling, with Channel 19 as a favourite with long-distance transport drivers.

UK AMATEUR RADIO FREQUENCIES (ITU REGION 1)

Band (metres)	Frequency limits (MHz)	Transmission modes
160	1.81 − 1.84	CW
	1.84 ± 2 kHz	RTTY
	1.84 − 2.00	CW and Phone
80	3.50 − 3.60	CW
	3.60 ± 20 kHz	RTTY
	3.60 − 3.80	CW and Phone
40	7.00 − 7.04	CW
	7.04 ± 5 kHz	RTTY
	7.04 − 7.10	CW and Phone
30	10.100 − 10.140	CW
	10.140 − 10.150	RTTY
20	14.00 − 14.10	CW
	14.07 − 14.099	RTTY
	14.10 − 14.35	CW and Phone
17	18.068 − 18.100	CW
	18.100 − 18.110	CW and RTTY
	18.110 − 18.168	CW and Phone
15	21.04 − 21.19	CW
	21.14 − 21.151	Reserved for international beacons
	21.151 − 21.450	CW and Phone
12	24.890 − 24.920	CW
	24.920 − 24.930	CW and RTTY
	24.930 − 24.990	CW and Phone
10	28.00 − 28.20	CW
	28.200 ± 50 kHz	CW and RTTY
	28.20 − 29.70	CW and Phone

VHF AND UHF AMATEUR BANDS ARE AT:

50.00 − 52.00 MHz
140.00 − 146.00 MHz
430.00 − 440.00 MHz

Other bands are taken from parts of the spectrum extending up to 47,200,000 MHz (i.e. 47.2 GHz)

AMATEUR NETS WORLDWIDE

Operating frequencies of amateur nets may change to avoid interference with other stations that may be operating close by. If you do not find the net you are looking for, try searching frequencies for a few kHz on either side. Operating times of nets may also change, to suit seasonal variations in traffic or other commitments of those people running them.

Name	Time (UTC)	Frequency MHz	Notes
UK Maritime	08.00 & 1800	14.303	A long-standing net covering the North Sea, Atlantic and Mediterranean. Weather discussions at 18.30.
Trans-Atlantic	13.00	21.400	Operates mainly during the Atlantic-crossing season.
Intermar	06.00 to 11.00	14.313	Operates daily but on week days, 0900–10.00, control is passed to the **Cyprus net**. 14.313 is a backbone frequency used by a chain of nets around the world. Operating times are arranged to take advantage of propagation changes that occur as the earth rotates. Areas covered are the Pacific Ocean, Indian Ocean, Arabian Gulf, Red Sea, Mediterranean Sea, Atlantic Ocean, Caribbean and the American Continent.
Australia New Zealand & Africa Net	05.00	21.200	Covers South Pacific and Indian Ocean.
Tony's Net	21.00	14.315	Covers the South Pacific.
Pitcairn net	06.30	14.180	Mondays only. Covers the South Pacific.

Frequency lists

Pitcairn net	17.00	28.950	Tuesdays only. Covers the South Pacific.
Pitcairn net	22.00	21.350	Tuesdays only. Covers the South Pacific.
Pitcairn net	16.30	21.350	Fridays only. Covers the South Pacific.
Traveller's net	03.00	14.116	Covers the Indian Ocean.
South African	06.30	14.320	Covers the Indian Ocean.
South African	11.30	14.105	Covers the Indian Ocean.
South African	11.30	14.316	Covers the Indian Ocean.
Caribbean net	07.27	11.00	Covers Caribbean
Waterways net	11.45	7.628	Covers the US East Coast waterways and Caribbean.
Bay of Islands	19.00	14.329	Covers South Pacific, Australia and New Zealand.
Bay of Islands	07.15	3.820	Covers South Pacific, Australia and New Zealand.
Colin's net **Round Table**	17.00	14.175	A long-standing DX net which operates daily.

III. Weather information

WEATHER BROADCASTS

For several decades, English-speaking seamen operating in Western European waters have depended upon the BBC Radio 4 shipping forecast as their main source of weather information. This is broadcast on 0.198 MHz and times are:

0033
0555
1355
1750

NB These are local UK times, *not* GMT or UTC.

With a reasonable receiver, broadcasts can be picked up in all areas covered, which extend from Iceland and Norway to the south-western Spanish coast. Forecasts always follow the same ordered format, which makes them easier to follow when reception is poor and is of help to those whose understanding of English is weak.

Many other local and national radio and television stations also transmit shipping forecasts between their normal programmes and perhaps the most certain way of obtaining up-to-date information on these is to check with newspapers and periodicals which publish details of forthcoming programmes.

VHF coastal stations are another good source of weather information and broadcasts are sometimes made in English in addition to the national language. Procedures differ and sometimes (as in Gibraltar, for example) this information is only available on request. At other stations it is broadcast at scheduled times. If you are in an unfamiliar area perhaps the simplest way of finding these details is to call in and ask.

LONG-RANGE HF FORECASTS

Many long-range HF stations use SSB, Morse or telex to transmit detailed forecasts covering large sea areas. Volume 3 of the *Admiralty List of Radio Signals* gives details of hundreds of these stations, but here there is room for only a few. Those included are popular amongst long-distance cruising boats.

Radio France International

Freq. (MHz)	Time (UTC)
6.175	11.40
15.300	For all
17.620	frequencies
21.645	

Forecast areas include the Atlantic between 28° and 50°N and 10° and 50°W.
Areas also extend from the African coast, the Cape Verde islands to the eastern Antilles.

Station	Call	Freq. (MHz)	Time (UTC)			
Roma naval	IDQ	0.110	0500			
	IDQ	0.119		0930	1800	2100
	IDQ	2.291	0500			2100
	IDQ2	4.280	0500		1800	2100
	IDQ3	6.3903	0500	0930	1800	2100
	IDQ4	8.4846		0930		

Forecasts are given in English and cover the whole Mediterranean.

United Kingdom

Station	Call	Freq. (MHz)			Time (UTC)					
Portishead	GKA2	4.286	01.30	05.30	07.30	*09.30	†11.30	13.30	17.30	*21.30
	GKA4	8.5459	01.30	05.30	07.30	*09.30	†11.30	13.30	17.30	*21.30
	GKA5	12.822	01.30	05.30	07.30	*09.30	†11.30	13.30	17.30	*21.30
	GKA6	17.0984	01.30	05.30	07.30	*09.30	†11.30	13.30	17.30	*21.30
	GKA7	22.467	01.30	05.30	07.30		†11.30	13.30	17.30	

* Bulletins broadcast at these times are for the following areas:

Part 1 – East Northern Section – 55° to 65°N; 15°   to 27.5°W
Part 2 – West Northern Section – 55° to 65°N; 27.5° to 40°W
Part 3 – East Central Section   – 45° to 55°N; 15°   to 27.5°W
Part 5 – East Southern Section – 35° to 45°N; 15°   to 27.5°W
Part 6 – West Southern Section – 35° to 45°N; 27.5° to 40°W

† Bulletins broadcast at this time are for:

Part 4 – West Central Section   – 45° to 55°N; 27.5° to 40°W

USA

Station	Call	Freq. (MHz)	Times (UTC) and comments

Portsmouth (US Coast Guard)	NMN	4.4287
		6.6064
		8.7654
		13.1132
		17.3073
			04.00
			04.00 11.30 23.30
			04.00 11.30 17.30 23.30
			          11.30 17.30 23.30
			                      17.30
			Forecasts cover the Atlantic, north of 3° north and west of 35° west.

Ocean Gate (New Jersey)	WOO	4.387 8.749	Forecasts are transmitted at 12.00 and 22.00 UTC and are for the North Atlantic, west of 35° west, including the Caribbean and Gulf of Mexico.
Dixon (California)	KMI	4.402 13.083	Forecasts are transmitted 00.00 and 12.00 UTC. North Pacific areas covered are: Ox to 30°N and east of 140°W also north of 30°N and east 160°E.

Forecasts in Morse

Morse forecasts are transmitted at speeds of around 20 words per minute but, even if you have no decoder and are unable to read Morse at this speed, useful information can still be extracted if you record the forecast and replay it slowly. Though this may be tedious at first, most forecasts follow a routine format and, once you are aware of this, decoding by ear becomes easier. With a little perseverance it is often possible to pick out only the information of interest and ignore the bulk of the transmission.

Italy

Station	Call	Freq. (MHz)	Time (GMT/UTC)					
Roma	IAR25	0.519	00.50	06.50		08.30	12.50	18.50	20.30
	IAR24	4.292	00.50	06.50		08.30	12.50	18.50	20.30
	IAR28	8.530			07.00*	08.30		19.00*	20.30
	IAR23	13.011			07.00*	08.30		19.00*	20.30
	IAR27	17.160			07.00*	08.30		19.00*	20.30
	IAR22	22.3725			07.00*			19.00*	

* Forecasts for the whole Mediterranean are given at these times.

Forecasts are transmitted in Italian and English and give gale warnings, a synopsis and further 12-hour outlook for Italian waters between 6° east and 20° east.

WEFAX – FREQUENCIES AND SCHEDULES

Station	Call	Freq. (MHz)	Time (UTC/GMT)	Notes
Australia – Darwin				
	AZI32	5.755	23.00–11.00	Transmission schedule sent at 00.30
	AXI33	7.535	23.00–11.00	
	AXI34	10.555	24H	
	AXI35	15.615	11.00–23.00	
	AXI37	18.060	11.00–23.00	
Argentina – Buenos Aires				
	LRO69	5.185	24H	
	LRO72	10.720	24H	
	LRO84	18.093	24H	
Canada – Halifax				
	CFH	0.1225	24H	Transmission schedule sent at 10.14
		4.271	24H	
		6.330	24H	
		10.536	24H	
		13.510	10.00–22.00	
Germany – Offenbach/Pinneberg				
	DDH3	3.855		Transmission schedules sent on Mon. at 14.30
	DDK3	7.880		
	DDK6	13.8825		

India – New Delhi
	ATA55	4.9935	14.30–02.30
	ATP57	7.403	24H
	ATV65	14.842	24H
	ATU38	18.227	02.30–14.30

Italy – Rome
	IMB51	4.7775
	IMB55	8.1466
	IMB56	13.597

Japan – Tokyo
	JMH	3.6225
	JMH2	7.305
	JMH3	9.970
	JMH4	13.597
	JMH5	18.220
	JMH6	23.5229

Kenya Nairobi
	5YE1	9.0449
		10.115
	5YE3	17.3669
		2.2867

New Zealand – Auckland
	ZKLF	5.807	24H	Transmission
		9.459	24H	schedule sent
		13.550	24H	at 04.45 &
		16.340	24H	16.45

United Kingdom – Bracknell
	GFA21	3.2895	24H	
	GFA22	4.610	18.00–06.00	
	GFA23	8.040	24H	
	GFA24	11.0865	24H	
	GFA25	14.5825	06.00–18.00	
Northwood (RN)				
	GYA1	2.81385	16.30–07.30	30/9 – 31/3
	GZZ6	3.43685	19.30–04.00	1/4 – 29/3
			15.30–08.30	30/9 – 31/3
	GZZ2	4.24785	24H	
	GYJ3	6.43635	24H	

Weather information

	GZZ40	8.49485	24H	
	GZZ44	12.74185	24H	1/4–29/9
			07.30–16.30	30/9–31/3
	GYA61	16.93885	04.00–19.00	1/4–29/9
			08.30–15.30	30/9–31/3
	GYA	8.334	20.00–07.00	South Atlantic
	GYA	16.115	07.00–20.00	schedule sent at 15.00

United States – Honolulu (Hawaii)

	KVM	9.9806		
	KVM	11.0881		
	KVM	16.1331		
	KVM	23.3296		
Mobile				
	WLO	6.852	24H	Transmission
		9.1575	12.00–00.00	schedule sent on Mondays at 14.40
Norfolk				
	NAM	3.357	24H	Transmission
		8.080	24H	schedule sent
		10.865	24H	at 00.00
		16.410	09.00–21.00	
		20.015	12.00–21.00	
Pearl Harbor (USN)				
(Hawaii)	NPM	2.122	06.00–16.00	Transmission
	NPM	4.855	24H	schedule sent
	NPM	9.396	24H	on Wednesdays
	NPM	14.826	24H	& Saturdays at
	NPM	21.837	17.00–06.30	00.00 & 00.15

San Francisco
	NMC	4.3461	Transmission schedule sent at 20.19
		8.682	
		12.730	
		17.1513	

Senegal – Dakar
	6VY41	7.5875	20.00–08.30
	6VU73	13.6675	24H
	6VU79	19.750	08.30–20.00

South Africa – Pretoria
	ZRO5	4.014	17.30–03.00	Transmission schedule sent at 04.45
	ZRO2	7.508	24H	
	ZRO3	13.540	03.00–17.30	
	ZRO4	18.238	05.45–17.45	

Spain – Madrid
	ECA7	3.650	Transmission schedule sent at 18.10
	ECA7	6.9185	
	ECA7	10.025	

Rota (USN)
	AOK	4.704	18.00–06.00
	AOK	5.785	24H
	AOK	9.382	24H
	AOK	17.585	18.00–06.00

NAVTEX – STATIONS AND TIMES OF OPERATION

All transmissions are on 518 KHz and the modulation mode is SITOR B (FEC).

ID	Station	Call	Transmission times (UTC)								Duration
A	Antofagasta	CBA	00.00	04.00	08.00	12.00	16.00	20.00			10 mins
E	Magallanes	CBM	00.40	04.40	08.40	12.40	16.40	20.40			10
D	Puerto Montt	CBP	00.30	04.30	08.30	12.30	16.30	20.30			10
C	Talachuano	CBT	00.20	04.20	08.20	12.20	16.20	20.20			10
B	Valparaiso	CBV	01.00	07.00	13.00	19.00					5
R	Lisbon	CTV	02.50	06.50	10.50	14.50	18.50	22.50			10
	Montevideo	CWM	00.00	06.00	12.00	18.00					45
	Colonia	CWM	04.00	10.00	16.00	22.00					45
	La Paloma	CWM	02.00	08.00	14.00	20.00					45
	Punta del Este	CWM	05.00	11.00	17.00	23.00					45
	Salto	CWM	03.00	09.00	15.00	21.00					30
	Laguna del Sauce	CWM	05.00	11.00	17.00	23.00					30
F	Brest	FFU	01.18	05.18	09.18	13.18	17.18	21.18			15
G	Cullercoats	GCC	00.48	04.48N	08.48W	12.48N	16.48	20.48W			15
S	Niton	GNI	00.18	04.18W	08.18W	12.18	16.18N	20.18W			15
O	Portpatric	GPK	01.30	05.30N	09.30W	13.30	17.30N	21.30W			15
G	Dammam	HZG	06.05		12.05	18.05					15
H	Jeddah	HZH	07.05		13.05	19.05					15
B	Bodoe	LGP	00.18	04.18	09.00	12.18W	16.18	21.00			10
L	Rogaland	LGQ	01.48W	05.48	09.48W	13.48W	17.48	21.48W			10

NAVTEX – STATIONS AND TIMES OF OPERATION

All transmissions are on 518 KHz and the modulation mode is SITOR B (FEC).

ID	Station	Call	Transmission times (UTC)						Duration
V	Vardoe	LGV	03.00	07.00	11.00W	15.00	19.00	23.00	10
A	Buenos Aires	L2B	01.00	07.00	13.00	19.00			45
C	Bahia Blanca	L2I	03.00	09.00	15.00	21.00			45
B	Mar del Plata	L2P	02.00	08.00	14.00	20.00			45
D	Comodoro Rivadavia	L2W	04.00	10.00	16.00	22.00			45
E	Rio Gallegos	L3D	05.00	11.00	17.00	23.00			45
A	Miami	NMA	00.00	06.00	12.00	18.00			30
F	Boston	NMF	05.00	11.00	17.00	23.00			30
G	New Orleans	NMG	03.00	09.00	15.00	21.00			30
N	Portsmouth	NMN	01.30	07.30	13.30	19.30			30
R	San Juan	NMR	04.20	10.20	16.20	22.20			30
T	Oostende	OST	02.48	06.48W	10.48	14.48	18.48W	22.48	7
P	Scheveningen	PCH	03.48	07.48	11.48I	15.48	19.48	23.48	5
H	Häernösand	SAH	00.00	04.00	08.00W	12.00I	16.00	20.00	15
J	Stockholm	SDJ	03.30	07.30W	11.30I	15.30	19.30W	23.30	15
E	Samsun	TAF	00.40	04.40	08.40	12.40	16.40	21.40	
D	Istanbul	TAH	00.30	04.30	08.30	12.30	16.30	20.30	
F	Antilya	TAL	00.50	04.50	08.50	12.50	16.50	20.50	
I	Izmir	TAN	01.20	05.20	09.20	13.20	17.20	21.20	
R	Reykjavik	TFA	03.18	07.18	11.18	15.18	19.18	23.18	12

Weather information

P Petropavl	UBE	03.00	07.00	11.00	15.00	19.00	23.00	30
B Zhdanov	UDC	02.18	05.18	08.18	11.18	14.18	17.18	30
		20.18	23.18					
C Odessa	UDE	00.18	03.18	06.18	09.18	12.18	15.18	30
		18.18	21.18					
A Batumi	UFA	04.18	07.18	10.18	13.18	16.18	19.18	30
		22.18						
C Kholmsk	UFO	01.30	04.30	07.30	10.30	13.30	16.30	30
		19.30	22.30					
F Arkhangelsk	UGE	02.00	06.00	10.00	14.00	18.00	22.00	30
A Vladivostok	UIK	00.00	04.00	08.00	12.00	16.00	20.00	30
C Murmansk	UMN	01.20	05.20	09.20	13.20	17.20	21.20	30
U Tallinn	UNC	00.30	04.30	08.30W	12.30I	16.30	20.30W	30
Y Split	YUS	07.00	13.00		19.00			10
M Nicosia	5BA	02.00	06.00	10.00	14.00	18.00	22.00	10

A letter following the transmission times means the following:

I – Ice warning
N – NAVAREA warning
W – Routine area weather report

VOLMET

VOLMET broadcasts provide aircraft with reports of actual weather conditions at airports. Some of the information (e.g. cloud cover at various altitudes) is not likely to be of much interest to mariners but surface wind speeds and directions are. If there is an airport in the area in which you are sailing, it is often interesting to compare their weather conditions with your own.

VOLMET area				Frequency (MHz)				
African	2.860	3.404	5.499	6.538	8.852	10.057	13.261	
Caribbean	2.950	5.580	11.315					
Europe	2.998	3.413	5.505	6.580	8.957	11.378	13.264	
Middle East	2.956	5.589	8.945	11.393				
North Atlantic	2.905	3.485	5.592	6.604	8.870	10.051	13.270	13.276
Pacific	2.863	6.679	8.828	13.282				
South America	2.881	5.601	10.087	13.279				
South-east Asia	2.965	3.458	5.673	6.676	8.849	11.387	13.285	
Royal Air Force	4.722	11.200						

IV Technical data

DESIGNATION OF RADIO EMISSIONS

In 1979 the World Administrative Radio Conference agreed to a new designation system for radio transmissions. The following list is a selection of designations used to describe some of the more common types of transmission.

Morse

A1A Hand Morse sent by on/off keying of the carrier.

Telephony (speech)

A3E Amplitude modulation
J3E Single side band, suppressed carrier
R3E Single side band, reduced carrier
H3E Single side band, full carrier
F3E Frequency modulation

RTTY/AMTOR

F1B Direct frequency shift keying of the carrier
F2B Frequency shift keyed audio tone (FM transmitter)
J2B Frequency shift keyed audio tone (SSB transmitter)

Packet/data transmissions

F1D Direct frequency shift keying of the carrier
F2D Frequency shift keyed audio tone (FM transmitter)
J2D Frequency shift keyed audio tone (SSB transmitter)

BATTERY DATA

Battery-type cross reference

Volts	ITC No.	Can style	Zinc carbon Everready Blue seal	Silver seal	Alkaline Everready	Duracell
1.5	LR03	AAA	–	RO3S	LR03	MN2400
1.5	LR6	AA	R6B	R6S	LR6	MN1500
1.5	LR14	C	R14B	R14S	LR16	MN1400
1.5	LR20	D	R20B	R20S	LR20	MN1300
9	6LA61	PP3	PP3B	PP3S	6LF22	MN1604
9	–	PP9	–	–	–	–

(ITC – International Technical Commission)

Rechargable nickel/cadmium cell (sintered cell) data

Nominal voltage	Discharged voltage	Can style	Nominal capacity (amp hrs)*	Maximum continuous charge*
1.25	1.0	AAA	0.18	20mA
1.25	1.0	AA	0.5	66mA
1.25	1.0	C	2.2	250mA
1.25	1.0	D	4.0	500mA
8.4	7.0	PP3	0.11	100mA
8.4	7.0	PP9	1.2	120mA

NB *These figures are for guidance only. Capacities and maximum charge rates vary between manufacturers.

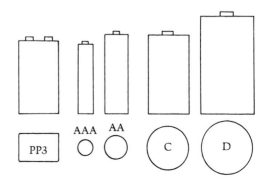

Fig. (i) Relative dimensions of various battery types

Further information

BOOK LIST

Admiralty List of Radio Signals, Vols. 1–6, published by the Hydrographer of the Navy. (This UK publication provides an authoritative source of world-wide radio information. It is kept up to date by weekly editions of Admiralty Notices to Mariners.)

Vol. 1 Coast radio stations; radio medical advice; arrangements for quarantine reports; locust reports; pollution reports; INMARSAT, maritime satellite service; regulations for use of radios in territorial waters; distress, search and rescue procedures; radio reporting systems and a brief extract from the International Radio Regulations.

Part 1 covers Europe, Africa and Asia (excluding the Philippines and Indonesia).

Part 2 covers the Philippine Islands, Indonesia, Australasia, the Americas, Greenland and Iceland.

Vol. 2 Radio beacons (including aero radio beacons in coastal regions); radio direction finding stations; coast radio stations giving a QTG service; calibration stations (i.e. stations giving special transmissions for the calibration of ships DF) and radar beacons (Racons and Ramarks).

Vol. 2a contains diagrams only, for use in conjunction with Vol.2.

Vol. 3 Radio weather services and related information, including certain meteorological codes provided for the use of shipping.

Vol. 3a contains diagrams only, for use in conjunction with Vols 1, 3, and 4.

Vol. 4 List of meteorological observation stations.

Vol. 5 Standard times; radio time signals; radio navigational warnings (including relevant codes and practices, ice reports);

and electronic position-fixing systems (Loran, Consol, Decca, Omega, differential Omega and satellite navigation).

Vol. 5a contains diagrams only, for use with Vols 1 and 5.

Vol. 6 Stations working in the Port Operations and Information Services; services to assist vessels requiring pilots; and services concerned with traffic management.

Part 1 covers NW Europe and the Mediterranean.

Part 2 covers Africa and Asia (excluding Mediterranean coasts), Australasia, the Americas, Greenland and Iceland.

Vol. 6a contains diagrams only, for use in conjunction with Vol. 6.

Guide to Utility Stations, Klingenfuss Publications, Hagenloher Str. 14, D-7400 Tuebingen, Germany. (An annual publication giving frequencies and schedules of many different types of station (except broadcast) world-wide.)

World Radio TV Handbook (WRTH), published by World Radio, 71 Bleak Street, London W1R 3LF. (An annual publication giving frequency lists and operating times for broadcast stations world-wide. Also contains much other useful associated information.)

The Handbook for Radio Operators, UK Post Office copyright, published by HMSO. (Intended to provide guidance for radio operators using frequencies within the maritime mobile bands.)

Maritime Radio Services Guide, published by British Telecom International. (Available to users of commercial radio services. It includes lists of frequencies charges, operational details and a description of services provided by British Telecom on VHF, MF and HF.)

Radio Amateurs' Examination Manual by G. L. Benbow, published by the Radio Society of Great Britain. (An examination course book.)

How to Become a Radio Amateur, published by Department of Trade and Industry. (Gives a comprehensive explanation of licensing requirements. Information sheets on specific aspects are also published.)

Radio Data Book by G. R. Jessop, published by the Radio Society of Great Britain, 1985. (Packed with technical information on circuit design, antennas, engineering data and many other topics.)

Practical Wire Antennas by John D. Heys, published by Radio Society of Great Britain, 1989. (Describes many simple types of wire antennas with a minimum of theory.)

Amateur Radio Software by John Morris, published by the Radio

Society of Great Britain. (Gives a full description and BASIC and machine code listings for a variety of radio-related programs. Includes CW and RTTY.)

Marine Electronic Navigation by Appleyard, Linford & Yarwood, published by Routledge & Kegan Paul, 1988. (A very comprehensive work, including in-depth information on most aspects.)

Piloting with Electronics by Luke Melton, published by International Marine Publishing. (Contains chapters on knot meters, echo sounders, Loran-C and piloting.

The 12-Volt Doctor's Practical Handbook by Edgar J. Beyan, published by Spa Creek Instruments Co., Annapolis, Maryland, USA, 1983. (A comprehensive and thoughtful guide to small-boat electrical systems.)

Waveguide Radio Receivers' Reviews, a BBC Publication, available from: BBC World Service Shop, Bush House, London WC2. (A set of receiver reviews conducted with the needs of short-wave broadcast station listeners in mind.)

VHF Radio Telephony for Yachtsmen, published by the RYA. (A concise paperback covering VHF procedures and operation. A useful primer for the Restricted (VHF only) RT examination.)

Weather Forecasts, published by the RYA. (Another useful paperback, revised annually, including sections on: understanding forecasts, forecasting, forecast frequency lists, a 7-language vocabulary and charts defining forecast areas used in several European countries.)

Contact addresses

The Radio Society of Great Britain, Lambda House, Cranbourne Road, Potters Bar, Herts EN6 5JE, UK

The Radio Amateur Licensing Unit, Chetwynd House, Chesterfield, Derbyshire S49 1PF, UK

Royal Yachting Association, RYA House, Romsey Road, Eastleigh, Hants SO5 4YA, UK

International Maritime Satellite Organization, 40 Melton Street, London NW1 2EQ, UK

Department of Trade & Industry, Radio Communications Division, Licensing Branch (Marine), Waterloo Bridge House, Waterloo Road, London SE1 8UA, UK

The Maritime Radio Examinations Group, British Telecom International, (Also Assistance Maritime and Aeronautical), 43 Bartholomew Close, London EC1A 7HP, UK

Index

antenna
  active 119
  feeders 106, 114
  half wave dipole 104–110
  height 71
  matching 111
  random wire 111
  testing 116
  tuner 73, 74, 111–113
  whip 115
    separation between nav. lights 117
alkaline cells 82
amateur
  frequency band plan 143
  nets 144–145
Amplitude Modulation (AM) 16
AMTOR/SITOR 53–57
authority to operate 21, 22
autolink 71
Automatic Repeat Request (ARQ) 54
ASCII 52–53

bandwidth 16
batteries
  data 158
  for portable equipment 81
  life 65
beat frequency oscillator 65
broadcast bands 130
  stations 131–133
bulletin board 62

call signs 22, 38
capacitors 90, 92, 94

capture effect 26
channels 24
charging of batteries 83, 84, 85
chokes 90, 92
Citizen Band (CB) 10, 21, 23, 26
Clerk-Maxwell 5
coastal station 28
coaxial cable 96, 106
command code 49
competence, certificate of 21, 22
computer
  connection to a radio 44
  interface 43
  requirements 42
  software 43, 46
current requirements 42, 87, 88

dellinger fade out 13
digipeating 61
digital selective calling/dialling 71
direct conversion receiver 77
dummy load 77, 118
duplex channels 25, 73, 74

earthing plates 100
electrons 10
EPIRBs 22
  batteries for 82
examinations 21, 36

fading (ionospheric) 10–14
ferrite
  beads 94, 95
  rods 69, 92
  toroids 94

Index

'FM'-ing 98
Forward Error Correction (FEC) 54
frequency
  coverage 66
  of oscillator 78
  relationship with wavelength 5
  selection 41, 70, 74
  stability 67
Frequency Modulation (FM) 18–19

gassing (of batteries) 83
grounding (RF) 99, 116

Hertz 4, 6

impedance of coax 106
INMARSAT 28
installation of HF transceivers 98
insulators 110
interference
  computer generated 42
  connected 90
  on board sources 88–90
  battery chargers 90
  to FAX receivers 88
  radiated 93
  to other radio and TV users 95–96
intermediate frequency filter 74
International Maritime Organization (IMO) 55
ionospheric layering 10, 14
ITA No. 2 code 4, 52, 53, 54, 122

kits 77, 78, 79

LAMTOR 55, 57
lead/acid batteries 82, 86, 88
lightning 89, 96
lithium/thionyl chloride cells 82

Marconi 5
marine VHF 24, 70
marine MF & HF 26, 72
marine band on domestic receivers 66

MAYDAY signal 124
modulation 14, 19
  classification 157
Morse 3, 12, 27, 37
  character definition 2
  decoding 51
  dot and dash relationship 122
  transmitter 78
multi-mode data controller 43–51

NAVAREA 56
NAVTEX 22, 41, 55, 55–57
  schedules 153–155
nickel/cadmium cells 83, 85
noise bridge 77, 119

operator licences 20, 35
oscillator 78

packet radio 40, 49
PAN-PAN signal 124
phonetic alphabet 123–124
polarity (RTTY) 52, 61, 62
portable receivers 63, 69
power connections 98
primary cells 82
propagation
  predictions 13–14
prowords 125–126

Q code 52, 127–129

radio telephone channels 138, 139
range 14, 25
rechargable batteries 82
refraction, ionospheric 10
ripple, low frequency 90
RST code 129
RT certificate 21
RTTY 52, 53

safety signals 125
satellite communications 22, 28
Seatel 71
secondary cells 82
securité signal 125
Selcall number 23
selectivity 67
self discharge 85

sensitivity 67
serial port 42, 44, 45, 72
shelf life 82
shift (RTTY) 53
ship's licence 22
ship to ship channels 137, 139
shore power 85
simplex 25
single band receiver 77
Single Side Band (SSB) 17, 18, 65, 72, 75
solar flare 13
spark 93
   discharge path 97
   transmitters 4, 6
spectrum 9, 24
spectrum analyser 74
speech compressor 74
standing waves 104
   ratio 118
Sudden Ionospheric Disturbance (SID) 13
sunspots 10
superhet 78
switching times
   AMTOR/SITOR
      transmit/receive 42

telegraphy 1
telephony 16, 20
teleprinter codes 4, 122, 123
television licences 24
terminal voltage 83
thunder storms, see lightning

time signals 141–142
toroid 94
transient supressors 94
transportable ship's licence 23
tuning 41, 66
   WEFAX 43, 59
   antennas 118
type approval 20, 24, 70, 74

ultra violet 10, 13
urgency signal 124

velocity factor 107
VHF
   antennas 115
   channel allocations 139–140
volmet 156
voltage
   compatibility 85
   fluctuations 86, 98
   spikes 90

wavelength
   relationship with frequency 5
weather 146–155
   FAX (WEFAX) 43
   FAX frequencies and schedules 149–152

X rays 10, 13

zinc carbon cells 82

Zulu time xii